OMG!

Operations Management Game

A Customizable Serious Simulation Board Game
for Learning the Core Principles of Operations Management

Akhmad Hidayatno
Armand Omar Moeis
Hariyanto Salim
Diana Wangsa Heryanto

A Publication by

Systems Engineering, Modeling and Simulation Laboratory
Industrial Engineering Department
Faculty of Engineering
Universitas Indonesia
2013

Preface

It is our honor to introduce the Operations Management Game (OMG!), the first customizable serious simulation board game in Indonesia, or probably the world. OMG! is designed to help managers and experts learn core principles of operations management in a simulated production line. OMG! is an example of a serious simulation game (SSG), i.e. a game that represents a real-world system and is designed for the purpose of learning and problem-solving. In SSGs, participants are immersed in a simulated learning environment that is risk free yet challenging at the same time. This environment creates an ideal place for testing decisions and learning from feedback without the costly disruption of a real system.

OMG! is one of many SSG products developed at SEMS Lab - Systems Engineering, Modeling and Simulation Laboratory. One of our focus areas is building SSGs that deal with complex problems, such as supply chain management, industrial management, and technology policies. Some of our products are: the *kontainer* game, the financial management game, the BIEOND biodiesel industry policy simulator, the franchise management game, the reverse logistics simulator, and many others. Almost all of our SSGs were developed using various computer simulation technologies, which is part of our lab's core competency.

SEMS Lab is also involved in developing solutions for complex systemic problems using multi-approach multi-method modeling and simulation tools. In fact, SEMS Lab is the only research laboratory in Indonesia that has a complete set of licensed modeling and simulation applications for delivering specific solutions or combinations of modeling solutions. Research at our campus has enabled us to develop and collect a library of knowledge on what is the best way to develop simulation solutions.

In this book, we will start with the story of OMG! to give you an overview of the ideas behind the development game. Then we will discuss about the history and concepts of serious games and simulation gaming, followed by explanation of the basic concepts in operations management that we want the game will teach. The next chapter will explain in detail the components and gameplay of OMG!. In the final chapter, we give you an example of playing OMG! with 6 stations configuration.

We would like to express our gratitude towards our parents, family and friends for their continuous support. We also would like to thanks SEMS Lab Assistants, Researchers and Alumni for creating together a very conducive lab for learning and innovating.

Thank you for becoming part of our journey in developing serious simulation games. I invite you to share your ideas for the benefit of future SSG development. The colorful standard size templates, which you can redesign and use to play the OMG! is available to our website at systems.ie.ui.ac.id.

Akhmad Hidayatno
Armand Omar Moeis
Hariyanto Salim
Diana Wangsa Heryanto

Contents

Preface 2

Contents 4

1. **The Story Of OMG!** 7
   1.1. Learning The Essentials 8
   1.2. Fun and Engaging Learning 10
   1.3. Customization 15
   1.4. Easy To Reproduce 16
   1.5. Who Is It For? 16

2. **Simulation Gaming and Serious Games** 17
   2.1. Where It All Started 17
      2.1.1. What Are Serious Games? 19
      2.1.2. Is Using Games For Non-Entertainment Purposes Really New? 19
   2.2. What Are Simulation Games And What Are They Good For? 22
      2.2.1. Game Designers' Perspective 22
      2.2.2. Simulation Scientists' Perspective 23
      2.2.3. Conclusions From Simulation Games Description 23
      2.2.4. What Are They Good For? 23
   2.3. The Evolution Of Simulation Games 24
   2.4. Who Acknowledges the Serious Games Initiative? The Emergence of Serious Games 26
   2.5. What We Can Learn From Game Designers 27
   2.6. Current State of Simulation Gaming and Serious Games 27

3. **Operations Management Principles in OMG!** 29
   3.1. Operations Management 29
   3.2. The Elements 30
      3.2.1. Planning What You Want To Do 30
      3.2.2. Preparation Is Key 32

	3.2.3.	The Devil Is In The Execution 34
	3.2.4.	Change The Sail And Direct The Ship (Control)....................36
	3.2.5.	Seeing Is Believing (Monitoring)..38
	3.2.6.	Do It Better Next Time (Improvement) 39

4. Playing OMG! 41

4.1. Physical Components...41
- 4.1.1. Tablemat Board..41
- 4.1.2. Information Cards 42
- 4.1.3. Upgrade Cards...48
- 4.1.4. Materials..49
- 4.1.5. Combining The Elements On The Tablemat.........49
- 4.1.6. Record Forms..52
- 4.1.7. Dice Throws Or Random Number Spreadsheet....56

4.2. Key Concepts 57
- 4.2.1. Assignable Costs...57
- 4.2.2. Upgrading 58
- 4.2.3. Variability..59

4.3. Gameplay 60
- 4.3.1. Game Stages 60
- 4.3.2. Rules 61
- 4.3.3. Game Steps 61

4.4. Customizing OMG! 62
- 4.4.1. Customization of Product Variation 62
- 4.4.2. Design A – 5 stations, 3 parts and 2 products........62
- 4.4.3. Design B – 6 stations, 4 parts and 4 products 64
- 4.4.4. Design C – 8 station, 5 parts and 6 products 66
- 4.4.5. Customization of Upgrades.....................................67
- 4.4.6. Customization of Variability 68
- 4.4.7. Customization of Performance Variables 69

4.5. Recommended Steps in Customizing OMG! 70
- 4.5.1. Define Game Objectives 70
- 4.5.2. Translate Objectives Into Customization Variables, Constraints And Upgrades 70
- 4.5.3. Re-Design Components (Tablemat, Parts, Upgrade Cards, Forms) 71

4.6. Prototyping And Finalizing The Game 72

    4.7. Evaluation And Continuous Improvement 72

5. OMG! Example With 6 Stations 73
    5.1. Suggested Table Layout 74
    5.2. Materials For Physical Components 75
    5.3. Station Descriptions 76
        5.3.1. Purchasing Station 76
        5.3.2. Production Station (M1, M2, and M3) 76
        5.3.3. Assembly Station 77
        5.3.4. Sales Station 78
    5.4. Upgrade Cards 79
    5.5. Game Variables 80
    5.6. Record Forms 80
    5.7. Stages And Rules For 6-Stations Game 86
        5.7.1. Briefing 86
        5.7.2. Playing 87
        5.7.3. Debriefing 87
    5.8. Special Notes for Playing in a Multi-Team Environment 88

Readings 89

1. THE STORY OF OMG!

The idea for OMG! (Operations Management Game) came while exploring the use of serious simulation games (SSGs)[1] to teach the basic principles of operations management. We were inspired by the famous MIT Beer Game, which is used for teaching supply chain management and systems thinking.

Our lab has actually developed several computer-based SSGs for various learning purposes in operations and production management, such as a beer crate production simulation game and a reverse logistics simulation game. However, tabletop games create a different learning atmosphere, where participants can touch the game and talk to each other during playing. The dynamic of conversation, debate and mockery while playing the game in a group setting is very interesting.

It is a challenge to create a physical tabletop game. It should be simple enough considering the limited space of a tabletop and challenging enough to keep the participants motivated to play the game and learn. As Leonardo da Vinci said, "Simplicity is the ultimate sophistication."

We decided that OMG! should represent the basic operational processes of demand fulfillment, raw material procurement, production through machining and assembling, and finally sale of final products based on demand.

In developing the game, the requirements for the design were:
- Customizable, reflecting different production sequence flows
- Capturing the essentials, yet providing the ability to introduce complexity in operations decisions
- A fun and engaging learning experience
- Easy to reproduce with off-the-shelf materials

---

[1] In this book, we decided to use the term serious simulation games (SSGs), to replace the term 'serious games and/or simulation gaming'. This is to bridge a widely perception in public, especially in Indonesia, that game is not suitable for learning,

1.1. Learning The Essentials

OMG! was born from the idea that learning should be fun and engaging, also the challenge of learning how to make operations decisions. The game is not meant to replace lectures or other more traditional methods of learning – we see it as a complementary tool to enhance the experience of learning.

As a complimentary tool, OMG! does not have to teach everything about operations management theory. It focuses on illustrating how changes in one variable are interlinked with changes in other variables.

We distinguish four basic management activities:
- Planning
- Organizing
- Executing
- Control

In *planning*, we distinguish two levels: strategic planning and production planning. Strategic planning requires a deep understanding of how the production and supply chain (material flow) work together in order to make decisions about investment in resources (man, money, machines, materials and methods). In production planning you must take variability in the system into account in order to develop a realistic production plan.

In *organizing*, the most important aspect is the activity of continuous improvement. Continuous improvement is usually sparked by competition with other companies. Competition with companies that offer similar products provides the drive to improve continuously.

In *execution* and *control*, information flows are critical for giving insight into what went wrong, what went right, and adjusting your direction accordingly. Money flows are also among the primary indicators in execution and control, since they show how efficient and effective your operations are.

In any serious simulation game, for learning to occur and surpassing the experience of merely playing a game, there are three stage of activities: *briefing, playing* and *de-briefing*.

1. Briefing

    The briefing should cover the purpose of the game, winning criteria, rules and gameplay. Facilitators can present some theoretical information for participants who come from a different background and to refresh the other participants' knowledge on the subject. Leveling the playing field ensures that the game can be played competitively in a multi-team setting. During the briefing, animated presentations or videos of the gameplay can help the participants prepare for playing. In the case of a complex game, the participants should always go through a couple of test runs to familiarize themselves with the gameplay.

2. Playing

    During gameplay, the participants must have the feeling that they are in control of their decisions and activities inside the game without further instructions. They should be encouraged to do what they are used to doing and not try to do anything differently. It is through mistakes and success in playing the game that the learning process is effective.

3. Debriefing

    The debriefing is the most important part of a serious simulation game. The debriefing should consist of a Q&A session, a discussion and a questionnaire. This is the stage of learning, brainstorming, and evaluating the experience that was gained during gameplay.

    The debriefing can consist of:
    - Process analysis of the game and what the inputs and outputs were of each process.
    - Clarifying assumptions and facts when making decisions, and how they relate to the concepts and principles the participants already knew.
    - Identifying the participants' different views and arguments regarding their experience and process interpretation.

- Identifying the participants' emotional experience during gameplay.
- Identifying the experience that the individual participants and the teams as a whole have gained

1.2. Fun and Engaging Learning

OMG! as a serious simulation game is a combination of gaming, simulation and interactive case-study. The differences between the three are listed in Table 1-1.

Game means interaction between players within a certain setting and under certain conditions, based on rules and procedures. Games are a place where players make choices, implement them and accept the consequences in order to achieve goals. Games are also fun, because they involve the players in a certain role, interacting with their environment in actual or simulated competition. They must overcome challenges, riddles, puzzles, and more, often in competition with other players. However, learning in game activities usually focuses on the game rules and the set of actions needed to win the game.

TABLE 1-1 GAME, SIMULATION AND CASE STUDY ACTIVITIES

Game Activities	Simulation Activities	Interactive Case Study
• Competition between players, groups or with the computer	• Real-world representation	• Detailed study of problem situation
• Rules of the game	• Risk free in simulated time (faster than real time)	• Explicit purpose by investigating a specific parameter or characteristic of the case

Simulation uses simulated conditions to test different results from a model, assuming the results would not significantly differ from results in the real world. A model is a real-world representation that only needs a limited number of characteristics of the real world. The modeling process must be conducted in a certain way in order for the simulation results to be acceptable. Learning in simulations mostly focuses on what-if decisions, which can be unlimited. There is no story. The players are only encouraged to change combinations of variables in view of certain goals.

One of the primary strengths of a simulation game is that it runs in a virtual world. This virtual world has its own time, which is not equal to real time. This can make decision feedback come up almost instantly, which is ideal in a learning environment. Feedback that would take three years in the real world, could take only three minutes in a simulation game.

A virtual world also provides the opportunity to simulate conditions that rarely happen in real life, but for which specific critical knowledge and skills must be learned in case they *do* occur (Baker, Navarro, & van der Hoek, 2003). A flight simulator for airline pilots, for example, is used for practicing emergency conditions during take-off, cruising and landing. Even though the majority of pilots will never experience any of these situations in the real world, they still have to practice their skills in handling them.

Another strength of simulation is that it provides a risk-free environment, because in it you are not dealing with the real world. This risk-free environment can foster a better learning process, because the decision-maker can test different sets of decisions without incurring real world costs if a decision turns out bad. One decision can be tested in different environments, or different decisions can be tested in the same environment. The absence of real risk allows participants to increase their confidence level in a less stressful but still stimulating environment (Alinier, 2003).

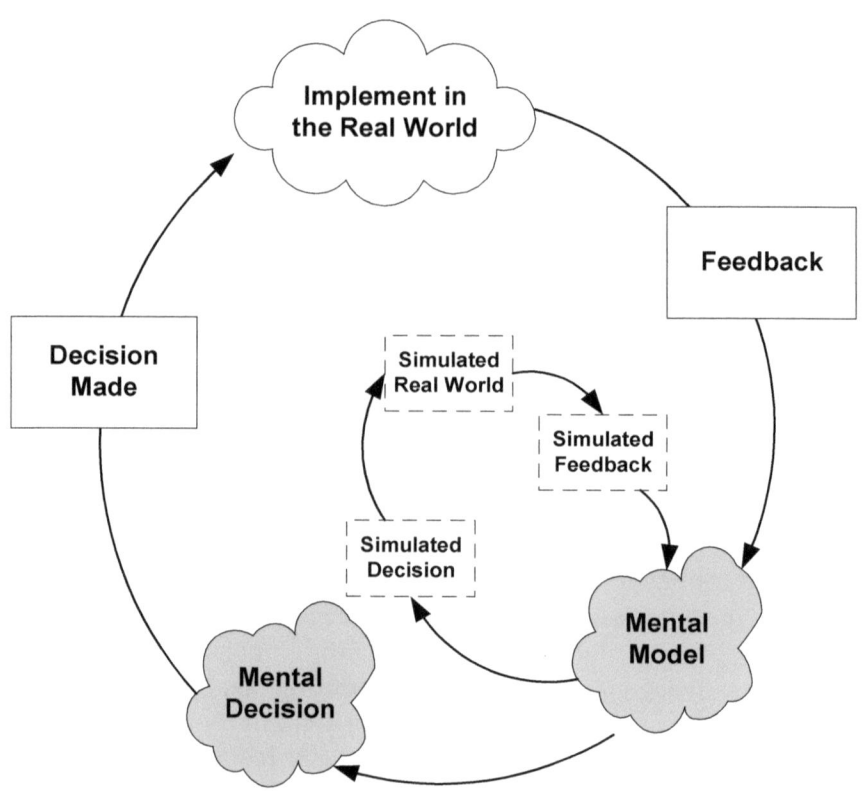

FIGURE 1-1 LEARNING LOOP USING A VIRTUAL ENVIROMENT FOR TESTING DECISION-MAKING (STERMAN, 2000)

These two primary strengths of the virtual game world make it an ideal platform for decision simulators or flight management simulators (Sterman, 2000). It avoids the costs of trial-and-error loops in the real world, replacing it with a simulated learning environment, as illustrated in Figure 1-1. Simulation-based tools allow participants to develop a wider perspective, to connect with real-world situations and to deal with the realities of competitive business (Haapasalo & Hyvönen, 2001).

A case study is a condensed problem definition of a real-world case, complete with data for analysis. An interactive case study is a staged case study where problems develop based on choices that are made. Case-study learning is usually focused on a predetermined solution that is limited to the scope of the case.

Combining these three types of medium for learning creates a new medium with different stages that combines the strengths of each original medium, as listed in Table 1-2.

TABLE 1-2 TRIPLE COMBINATION OF GAME, SIMULATION AND CASE STUDY

Case Study	Simulation	Games	*This triple combination offers complexity and challenge as the best learning medium*
Specific knowledge development by analyzing a set of similar data	Games based on simulation (characteristic similar games)	Limited time, rules, orientation, judgment, result-oriented, win or lose	**Games**
Real-world case representation, non-competitive, high-skill transfer	Open-ended, behavior-oriented, real-world reflection focus, process-oriented		**Simulation**
Deep analysis, detailed orientation, decision-impact oriented			**Case Study**

Simulation game means a combination of game components (competition, teamwork, rules, participation and role-play) and simulation components (models of the real world) (Riis & International Federation for Information Processing., 1995). For a game to be considered a simulation game it must involve an empirical model of the real world.

Learning through interactive case studies in simulated games is a method that combines a simulation game with the specific problem-solving effort of a case study. Case studies offer a realistic learning experience, because they are, by nature, based on experience.

Mimicking the real world in case studies through challenges, stories and scenarios, can create a learning process that simulates an actual experience. This is called experiential learning. At a certain level, people cannot differentiate between an experience they got from a real situation and an experience they got from a simulated situation. Watching a sad movie, which is fictional, can trigger an emotional reaction similar to experiencing a sad moment in real life. The power of serious simulation games is that they combine three methods for learning, as illustrated in Figure 1-2.

In addition, simulation-based games played on a computer are great learning support tools for today's generation of university students, who have never experienced a world without computers. Many of them have spent a lot of time playing computer games and have learnt to apply complex sets of rules through game playing. Hence, the learning style of the Virtual Generation (V-gen) is very different from that of former generations (Proserpio & Gioia, 2007).

1.3. CUSTOMIZATION

Different operation systems behave differently depending on the structure of the interconnected variables and the different strengths of the interconnections. This condition also applies to the real world. Different production systems, however similar, can produce different behaviors. Therefore, it must be possible to adapt the game to each different production system.

The adaptability of OMG! should cover:
- Different production flows
- Different numbers of stations
- Different numbers of products and components
- Multi-level assembly processes
- Variability in capacity, costs and constraints
- Upgradability of various variables during operation

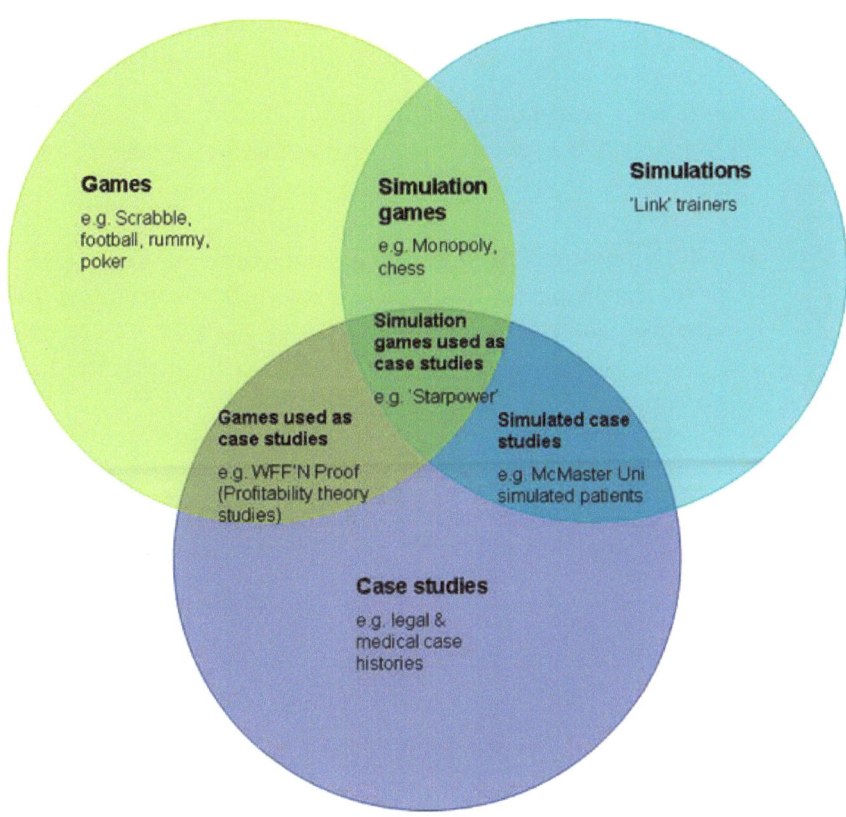

FIGURE 1-2 VENN DIAGRAM OF COMBINATIONS (LEIGH & KINDER, 2001)

1.4. EASY TO REPRODUCE

OMG! must be easy to reproduce using only off-the-shelf materials. The game board can be made with paper and color prints. The product components can be made with a combination of paper, stickers, paper clips, Lego bricks, or other materials, depending on how many product variations you decide to have. The product components should be light, small, and sufficiently visually distinct from each other.

In single-team play, OMG! uses a dice to emulate variability in each of the operation processes. However, in multiple-team play, OMG! uses printed or projected numbers that are generated using normal distribution. Therefore, the teams will all deal with the same variability. You can use any spreadsheet application to generate the numbers.

1.5. Who Is It For?

OMG! is built as a learning medium, which means the players are expected to have gained new insights after playing the game.

OMG! is for:
- Students learning operations management. With OMG! they can experience simulated real-world conditions and learn to make correct decisions.
- Managers or experts who need to understand their company's operational conditions.

2. SIMULATION GAMING AND SERIOUS GAMES

When confronted with the notion of games, most people think of *'entertainment artifacts'* that are used (played) to spend some leisure time, alone or with friends. Far from that narrow perspective, games are not limited to entertainment purposes at all. Since several decades many people have been captured by the potential of games and gaming for learning and other serious objectives, such as policy support.

2.1. WHERE IT ALL STARTED[2]

Since about the turn of the millennium, conceptual and technological developments in non-entertainment applications of gaming have reached a significant new level. Examples of these powerful changes in the field of gaming for advanced learning and other serious purposes can be found in the work of Marc Prensky and the Massachusetts Institute of Technology (MIT) Media Lab's Games-to-Teach project. Although the underlying concepts are still ambiguous, the initiatives in this field center around the exploration of digital game-based learning, simulation-based e-learning, serious games and serious play (Aldrich, 2004, 2005, 2009).

Mostly, the aforementioned projects and concepts are based on the awareness that more and more (young) people spend much of their leisure time playing computer/video games. As a result, many of the (computer) skills they acquire, but also much of what they learn about daily life, such as social interaction, is now unintentionally mediated through their experiences with computer and Internet games. For the present generation, these experiences start very early in life and are expected to continue until late in life.

In addition, it has not escaped the attention of researchers of economic, computer and social disciplines that the computer and video game industry is rapidly expanding its market and profits – it will soon

---

[2] Parts of this chapter were taken from Moeis, A. (2004). Online Serious Games for Infrastructure: a prototype and a set of recommendations, Master Thesis, Delft University of Technology, the Netherlands, August 18, 2004.

outperform the movie industry. At present, a few large game developers and publishers largely control the video gaming industry, while smaller companies are trying to find niche markets. One of the niche markets where some small companies have succeeded is that of educational games, e-learning and professional training.

For many teachers, pedagogical experts and curriculum developers, active learning is now considered one of the most effective learning styles to bridge the knowing-doing gap. The knowing-doing gap refers to a mismatch between what a person has learned and his or her abilities. The idea behind active learning is that the student or young professional learns by doing and from experience – in other words, by being provided an environment for hands-on experience and self-exploration.

It is clear that the aforementioned trends – changing paradigms of learning, the rapidly growing gaming market, and small innovative companies exploring and developing educational games – reinforce each other. Since a decade or more, this has mostly led to the application of games in education from kindergarten to high school, but the logical next step is the application of gaming in higher education and for professional lifelong learning and learning in a decision or policy making context.

Largely, the ideas, trends and notions introduced above have come together in the Serious Games Initiative. The Serious Games Initiative was formed from a loosely coupled network of researchers, game developers, technologists, and people working in consulting, computer industry, media, universities, government, etc., who are interested in the advancement and use of digital gaming for learning and policy support. This initiative has produced several working examples and with support from related institutions has held several official conferences on serious games (www.seriousgamessummit.com).

2.1.1. WHAT ARE SERIOUS GAMES?

What are serious games? Are they games that have serious content? What does 'serious' mean? Can we consider SIM CITY 5 an example of a serious game? Doesn't SIM CITY 5 actually have serious content, because in it we have to deal with problems like city budgeting, taxing, and spatial planning.

A serious game is any game whose prime mission is not entertainment. They do not include advertising games or games for K12 education (K12 education: kindergarten, elementary school, and high school), or any entertainment game that can be applied to a mission other than entertainment. Using this definition, we consider SimCity 4 a serious game, provided that it is used for non-entertainment and non-K12 education purposes.

2.1.2. IS USING GAMES FOR NON-ENTERTAINMENT PURPOSES REALLY NEW?

The use of manual or computer-supported games for learning, policy support and other serious purposes is not a recent innovation. Paper-based games such as Pank-A-Squith, published around 1910, have been used to address political issues – in this case the promotion of democratic rights of women (Figure 1-1).

The goal of this board game is to reach the House of Parliament, the main subject is voting rights for women. Although designed to be humorous, the images also evoke a darker side of what happened during those days, such as police violence against women protesters and force-feeding of imprisoned hunger strikers. Another early example is Sketches Of The Rebellion, published in Philadelphia by McFarland and Thomson in 1862, which was produced to encourage the Union cause during the American Civil War (Figure 2-2).

As can be seen from these examples, the use of games for non-entertainment purposes can be traced back to more than one century ago. Therefore, the idea of non-entertainment games is not new. What is new is the technology used to convey the serious games idea. The computer games industry has pioneered the use of sophisticated computer graphics technology for mass use. Computer games have become more realistic – not only the graphics, but also the game environments as a whole. Furthermore, human behavior in gameplay has also changed. People are spending more time playing games, especially computer and console games. People's interest in social interaction and online games has been proven to be substantial.

FIGURE 2-1 PANK-A-SQUITH (C. 1910)
(SOURCE: HTTP://RMC.LIBRARY.CORNELL.EDU/GAMES/INDEX.HTML)

2.2. WHAT ARE SIMULATION GAMES AND WHAT ARE THEY GOOD FOR?

Greenblat & Duke (1981) offer a definition of simulation games as: *a hybrid form of game, involving game activities in a simulated context*. Moreover, they stress that simulation games require human interventions, where the computer may serve as a high-speed calculator or it may contain a model/set of models that are triggered by the actions of the players (Greenblat & Duke, 1981). Although their definition can be used as a basis for understanding simulation games, we take it further by describing the meaning of each word – *simulation* and *game* – using definitions given by various publications about simulation science and game design. First we give a view from the perspective of the game designer and then from the simulation scientist.

FIGURE 2-2 SKETCHES OF THE REBELLION (1862)
(SOURCE: HTTP://RMC.LIBRARY.CORNELL.EDU/GAMES/INDEX.HTML)

2.2.1. GAME DESIGNERS' PERSPECTIVE

Two leading game designers, Salen & Zimmerman, define a game as: *a system in which players are involved in an artificial conflict, defined by rules, which results in a quantifiable outcome* (Salen & Zimmerman, 2003). They argue that this definition is applicable to any kind of game, whether it's a computer game, a board game, or a game like puzzling. Moreover, 'artificial' is defined as having a boundary from the real world, and 'conflict' is explained as a contest of powers (cooperation and competition). In another phrase, a game is: *an artificial system that provides its players with opportunities to compete or cooperate with each other (other players or simulated players) in achieving a goal.*

Salen & Zimmerman come up with the following definition of simulation: *a simulation is a procedural representation of aspects of "reality"*. Moreover, they argue that any game can be considered a simulation, but there are many simulations that are not games. Games like chess for instance, simulate a territorial war. Another example is Tetris, a game about managing falling objects that simulates the forces of gravity. What matters is not the means of the simulation (whether it uses a computer or a game board) but how it resembles an imaginative reality.

2.2.2. SIMULATION SCIENTISTS' PERSPECTIVE

Shannon defines simulation as: *the process of designing a model of a real system and conducting experiments with this model* (Shannon, 1975). Moreover, based on the perspective of system dynamics, simulation is a problem-solving tool that follows the changes over time in a dynamic system model. Although this view emphasizes the use of simulation as a problem-solving tool, Neelamkavil stresses that simulation models cannot find the optimum solution. What they *can* do is compare several alternative solutions (Neelamkavil, 1987).

What is the advantage of simulation? Pidd (1992) gives the following criteria: cheaper compared to real-system experimentation, time flexibility (possibility to go forward weeks/months in seconds), repeatable (can be replayed with the same settings), safety (can be used to simulate extreme conditions).

2.2.3. CONCLUSIONS FROM SIMULATION GAMES DESCRIPTION

We conclude that there are two things that game designers and engineers/scientists agree upon. First, simulation is a representation of reality. It doesn't have to be very close to reality, but simulation has to have some important characteristics of a real system. Second, Hoover & Perry (1969) share the same view with Salen & Zimmerman that games involve competition. Games should provide players the ability to compete with other players (or simulated players).

Combining Greenblat & Duke's definition of simulation games with the aforementioned description of games and simulation, we come up with two traits of simulation games. Firstly, human players trigger most of the events in simulation games. Secondly, simulation games are based on a model/set of models (which are played by human players).

2.2.4. WHAT ARE THEY GOOD FOR?

Greenblat & Duke describe how simulation games can serve as "the future's language." Duke mentions that gaming is necessary because human problems are becoming too complex (Duke, 1974). Traditional communication cannot bear the huge load of information that should be digested by human beings in order to preserve their race. Related to his background as an urban planner, Duke illustrates how urban centers are becoming more and more complex. This complexity cannot be managed with traditional forms of communication. Simulation games can serve as new forms of communication to solve this problem. They offer ways for people to have a dialogue; they offer ways for people to collaborate or even to confront each other.

Rhyne (in Duke, 1974) states that these complex problems have forced decision-makers to give attention to wider and more complex fields of knowledge. Knowledge has to be developed and distributed in an interdisciplinary approach. Today's problems cannot only be solved using one point of view. The one's advantage could be the other's disadvantage. Finding solutions requires multiple efforts and alternative approaches.

How can simulation games take up these challenges? To put it simple, simulation games can offer players an environment for experimenting with their actions. Players can see the impacts of their actions without having to bear real impacts. Simulation games provide a safe environment to implement a plan of action without having to interrupt a system in the real world.

We have to understand the difference between 'gaming simulation' and 'simulation game'. Gaming simulation refers to the theory that lies behind a product called a simulation game. Although both are nouns, 'gaming simulation' is used when we refer to a theory/concept, while 'simulation game' is used when we refer to a product (game).

In the next section, we present a brief description of internet-mediated simulation games. This serves as an introduction to further developments in simulation games with respect to technology and how it is used to improve games.

2.3. THE EVOLUTION OF SIMULATION GAMES

Simulation games started in paper-based form. Those first games consisted of game instructions and a paper-based model of the reality they represented. Later, computers were introduced to support calculations in simulation games (we label these *computer-supported simulation games*). At first, the computer was used as a standalone device that players or instructors could use to do some calculations. In the 1980s, LAN-based simulation-games emerged. Using the then sophisticated LAN technology (a LAN is a computer network covering a local area, such as an office or a home), simulation games were created for PC platforms. Players communicated and collaborated through a computer network. Being limited to the topology of the LAN, they were still playing in one local area, such as an office or a university campus. The latest form of simulation games are *internet-mediated simulation games*, which are characterized by having internet as the game platform. Thus, participants in different countries can play simulation games (in different time-zones, with different mother languages, etc.). Figure 2-3 shows an illustration of this evolution.

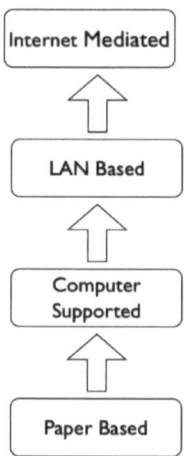

FIGURE 2-3 FROM SIMULATION GAMES TO SERIOUS GAMES

What we need in the future are better tools for research, learning, and intervention for infrastructure design and management. By combining the faculty's knowledge and experience in internet-mediated simulation games and simulation games for infrastructure with serious games, SEMS Lab hopes to come up with a new kind of tool that leads to better innovation in the future.

2.4. WHO ACKNOWLEDGES THE SERIOUS GAMES INITIATIVE? THE EMERGENCE OF SERIOUS GAMES

Ben Sawyer, with support from the Alfred P. Sloan Foundation, developed a game addressing university management called Virtual U (2000). The game was very successful, with a thousand copies sold and having been downloaded more than 15,000 times by the year 2004. With the success of this project, Sawyer teamed up with Dave Rejeski, director of the Woodrow Wilson International Center for Scholars' Foresight and Governance project, to initiate the first Serious Games Day in December 2003. This event was also used for the inauguration of the Serious Games Initiative.

In March 2004, at the Game Developers Conference initiated by the Serious Games Initiative, the two-day Serious Games Summit was held. The summit drew almost three hundred participants, most of them game developers and designers. The summit was supported by Microsoft Research.

The initial proponents of the Serious Games Initiative are the Woodrow Wilson' Foresight & Governance project and MIT's Education Arcade project. Other significant supporters are Stanford's Media X (http://mediax.stanford.edu) and Carnegie Mellon's Entertainment Technology Center (http://www.etc.cmu.edu). The initiative has also been acknowledged by the North American Simulation and Gaming Association (NASAGA, http://www.nasaga.org) and the Digital Games Research Association (DiGRA, http://www.digra.org).

The use of computer games to address non-entertainment/public-management problems started before the Serious Games Initiative was established. For example, SimHealth published by Maxis in 1993 is a healthcare policy management game. In this game, players have to balance the needs of communities, the medical industry, and insurance companies. Although SimHealth was developed as a commercial game, it has also been used to support college level education.

2.5. WHAT WE CAN LEARN FROM GAME DESIGNERS

Game designers have the skills to create captivating computer games. Those skills are supported by their capacity to design interfaces, their familiarity with designing 3D graphics and audio, developing artificial intelligence, characters and opponents, and online multiplayer gaming platforms. Game developers are not technology pioneers, but they do contribute to technology's usability by shoe-fitting cutting edge technology and programming practices into lower-end hardware.

Game developers have the skill to package or re-package models/simulations (simulation games) created by scientific-based model/simulation building communities. This can help simulation games reach their full potential. The serious games/computer games approach is not aimed at replacement but at enhancement of simulation games.

The Serious Game Initiative offers tools that can be used by researchers in gaming simulation to bring simulation games to the next level and gain a wider audience.

2.6. CURRENT STATE OF SIMULATION GAMING AND SERIOUS GAMES

In recent years, simulation gaming/serious games designers have been using game design technologies to the fullest. Most serious games are developed using 3D game design technologies. Serious games designers are working together with 3D animators, graphic designers, and game programmers to create games that were unimaginable before. This new breed of serious games uses the same technologies as commercial entertainment games. The subjects range from port planning and management to oil and gas safety management, urban planning and railway planning.

Simulation gaming/serious games have also reached a level where several games can be played worldwide. This development has surpassed what simulation gaming/serious games theorists have predicted in the past. From a technological point of view, it is now easy for a simulation builder/modeler to create a web-based simulation game using off-the-shelf web-based tools. Such tools are so robust that they enable a modeler to simply submit his model via a website to set up a web-based simulation in a matter of minutes. Adding multi-player interfaces requires minimum programming skills.

So how about the Old-Fashioned Paper-Based Simulation Games (Board Games)?

Actually, board games are still the most played simulation games. An example is the MIT Beer Game. It is still used in many system dynamics/system thinking related courses everywhere in the world. For complex games, most serious games designers still combine computer-based games with board games. Both complement each other and it seems that some things are still better done manually.

3. Operations Management Principles in OMG!

OMG! is a serious simulation game that was developed to help managers and experts learn various aspects of operations management in a simulated production environment. In this chapter, we will go into the principles of operations management that participants will learn to apply when playing the game.

3.1. Operations Management

In daily life, we all do operations management. We plan our daily activities and prepare for going to school or work. Then we execute our plans. We try to make sure that things go according to plan and control the aspects that we can control. We monitor the activities and observe the outcome and then improve the process from time to time by going through the cycle more efficiently.

Simply put, operations are a series of coordinated activities with certain goals at the end. We can divide operations into different stages:
- Planning
- Preparation
- Execution
- Control
- Monitoring
- Improvement

In a serious business environment, professionals manage operations. There is little room for error and the impact of operations management is directly reflected by the performance of the business in terms of financial results. Businesses that manage their operations better than their competitors perform better. Operations are usually unique for each type of business, but they always follow the logical sequence mentioned above.

In a manufacturing environment, operations management is important, if not everything. Professional managers forecast the demand in the market, and plan production capacity and prepare production schedules accordingly. They plan and release orders to suppliers to prepare the materials needed for production. During production, unexpected things can happen, from machine breakdown to not having enough components. The better operations are controlled, the less they are interrupted. Constant monitoring aided by statistical tools helps managers detect obstructions and respond appropriately. They do this repeatedly and get better from one cycle to the next.

In service industries, operations management is equally crucial. Being dependent on human resources they have to plan their manpower capacity carefully. They need to manage their relationship with their customers, they need to deliver what they have promised, and for that they need to prepare the right tools and equipment. They listen to their customers and monitor any other feedback. Service is improved on a regular basis, in respect of which employee training is very important.

3.2. THE ELEMENTS

As mentioned before, operations are a series of coordinated activities consisting of planning, preparation, execution, control, monitor and improvement. On the basis of some stories and examples we will go a little deeper into these separate elements to provide a better understanding of each.

3.2.1. PLANNING WHAT YOU WANT TO DO

Some things need a lot of planning, while the event itself may last only a short time.

Imagine you are going to get married. How important is the wedding party to you? How long in advance do you think you will start planning for it? How long will the party last? How many times do you think you will repeat your wedding party?

When you think about having a wedding party, you have actually already started planning. Some people plan their wedding party months or even a year ahead, while the party itself only lasts one day. Something that is so important has to be planned way ahead of the event itself.

Now imagine someone who has to plan, say, the Grammy Awards Nomination Evening Gala with all the VIPs and celebrities invited to the event. When do you think they start planning? And what about someone who wants to start up a business or manufacturing company?

Planning means anticipating events that will occur in the future. In operations management, planning comes before everything else. Your plans can still change of course, because the future is unpredictable. You also have to make backup plans, in case something unexpected happens that gets in the way of the original plan.

Planning differs from preparation, because during the planning phase there is no real action yet. Planning starts conceptually and is then put into text, schedules and blueprints that will become the basis for the activities called preparations. During planning you consider many available options and possibly options that aren't available yet. These are sorted according to priority and how likely they are to happen. You plan the best scenario with the available options for what is likely to happen.

When you want to set up a business, you prepare a business plan. This contains a projection of how you want the business to develop in terms of sales, cash flow, and investments needed to realize the plan. Of course, these projections are only forecasts. They serve to give you vision ahead of what might happen. A good business plan will have an optimistic, a normal and a pessimistic scenario, and numbers to help you prepare for the best and the worst. The quantitative aspect helps you determine what kind of preparations you need to make.

In operations management plans are reviewed periodically and altered when necessary. There are strategic plans, which stay the same for a long period of time. And there are medium-term and short-term plans, which are used for a shorter period of time.

Strategic plans include directives on how the company should operate, also called the 'vision and mission' of the company. They drive how the operations in a company evolve over time. They are valid for a long period of time and are only changed in a substantial way if the business environment of the company changes drastically.

Shareholders usually appoint a new chief executive officer (CEO) when the company suffers as a result of such changes and is forced to look for another direction. This is usually the only time when there is a change in the vision and mission of a company.

For a manufacturing company, the medium-term plans usually involve capacity planning. Demand forecasts are needed for establishing these plans, in order to calculate and establish the so-called 'rough cut capacity planning'. How many resources are needed to fulfill demand during a medium-long period, i.e. two to five years from now? This may mean hiring and preparing new workers, purchasing new machinery and equipment, or even building a new manufacturing facility.

In a manufacturing company, planning for materials needed is part of short-term planning. The process cycle of the material requirement planning (MRP) can be so short that some companies go through it daily. The MRP considers existing inventory, existing orders to suppliers, current and future production schedules, lead-times from suppliers, and receiving buffers. Subsequently, the component requirements are calculated. MRP produces a schedule for releasing orders to suppliers: how much should be ordered and when, so that you receive the components in the necessary quantity at the right time.

This kind of planning requires a certain amount of tolerance, especially when dealing with external factors such as external suppliers. This tolerance is related to the lead-time (the time a supplier needs to fulfill an order) and is implemented as a buffer, to prevent running out of components. The tolerance can be reduced somewhat when the lead-time is shorter. For example, if a manufacturing company produces its own components, it has more control of the lead-time.

3.2.2. Preparation Is Key

So you have planned everything – what's next?

The preparations are the set of actions that come after planning. Everything you planned should be prepared before the event. For a wedding party it can be as simple as making several phone calls to find the right venue and catering company, or as tedious as delivering invitations several weeks prior to the party. Like planning, preparations can take place a long time before the event itself and they can require a lot of effort.

A good preparation can help ensure that things go according to plan. Preparations are a test of your plan prior the event. During preparations you will sometimes find out if your plan is feasible or not, which enables you to change your plan accordingly. At this stage, you go back and forth between planning and preparation. The ultimate goal of preparation is to make sure the plan is perfected and can be carried out. During preparation, you can even decide to cancel the whole plan or event if you find out that it simply isn't feasible. Though cancelling can be costly at this stage, it is sometimes better to do it sooner than later.

During preparations, all parties involved are coordinated. You share information and seek support. When a business is started up, external financial support may be necessary. For this purpose, you need to share your business plan with potential investors. When new people are hired, you share information about your plans with the media, such as newspapers or social media, so that people can give support by applying as employees. Also, you need a place to run your business, so you coordinate with people that offer office and storage space (or with your parents and start up your business in the garage).

These preparations are all necessary before the business actually starts. You set up the infrastructure needed to pick up the first sales. Preparations are key to a successful operation.

Tactical planning decides how much capacity is required to fulfill future demand. Increasing or decreasing capacity can mean modifying the existing production facilities, including machinery, people, and layouts. Infrastructure can also be impacted, like changing to a new energy source, altering existing electrical layouts, expanding warehouses, etc. All of these preparations take place before the demand hits the company so that it has the capability to fulfill it. If you fail to realize everything on time, the company will be unable to cope with the demand and suffer from backlogs.

For short-term plans, such as the material requirement planning, preparation is even more critical due to the shorter period of time available. The MRP suggests which components to order from which suppliers and which to produce internally, at what time and in what quantity. These are merely suggestions to support the master production schedule (MPS). Actual orders to suppliers need to be placed according to the MPS. Also space to receive and store the components needs to be made available.

But what if the suppliers can't comply with your requests? What if there isn't enough space in your warehouse? These kinds of questions will come up during preparations, because you are dealing with real conditions that you may not have predicted during planning. Further action needs to be taken. If you have a multi-vendor system, you can switch to another supplier, re-balancing the orders among your vendors. If you adopted a sole-supplier strategy, the strategic thing would be to work together with your supplier to solve both his and your problems.

3.2.3. THE DEVIL IS IN THE EXECUTION

Planning and preparation are tested during execution. If they fail, execution will fail as well. If they are good, execution will most likely succeed.

You are like the conductor of an orchestra who decides at what precise moment the piano starts playing after the violin. If you do it right, the music will sound as written in the music score – harmonious and nice. If you fail, the music will not sound as nice and become a mess very quickly.

During execution, all the information sharing and coordination you did with other parties will come to a culmination point. Now, your job is to synchronize. During execution, you decide how one activity follows the other according to the plan and your preparations.

Tactical planning and execution to increase or decrease capacity is essential, especially in more established institutions. Your plan of getting more business can only be achieved if you match it with an increase in capacity. Your cost-saving plan can only be achieved if your capacity is reduced on time. Timing is essential during execution.

When setting up a business, you have to formulate your business plan carefully, make sure you have enough capital to buy from your suppliers, and prepare the capacity you need to deliver the products or services you are trying to sell.

What comes first: getting the market and sales or preparing capacity? Market first and build capacity immediately after. You have to time it carefully, so that when the first orders come in you have the capability and capacity to deliver. Remember that it takes time to build a market and get sales. It also takes time to build capability and capacity. Most importantly, it takes capital to create all of them. When you are starting up your business with limited capital, spend it on building capability not capacity. Try to grab the market and build capacity along with the size of the market you gain. When you have market and sales, but run out of capital, get a loan from a bank or find new investors.

Your preparations for getting resources, such as people, machinery and infrastructure, will have to be synchronized so that they don't disrupt running operations. If you have to shut down an assembly line, you want to have it going again as quickly as possible. Imagine you have already planned the assembly line to stop for three days starting tomorrow, but the machinery that was scheduled to arrive yesterday will not be there in another two days. With good preparation and planning you will know what's going to happen and can act accordingly. For example, you can postpone the assembly line stop for another three days and make sure that productivity is high before it is stopped.

Executing the master production schedule with the material requirement planning prepared beforehand is also crucial. You synchronize the timing of placing orders with your suppliers considering their lead-time so you receive the components on time. If the components arrive too early, you will have to store them in a warehouse. If you have a hundred components arriving one week early that require one pallet position each, you will need a hundred pallet positions in your warehouse. What if you handle thousands of components? How big do you want the warehouse to be?

You also don't want the components to be late or your assembly line will be forced to stop. Can you imagine all the resources sitting idle, yet still incurring costs? Of course you could choose to expedite the shipment using the most expensive delivery service that money can buy in order to get them on time.

If your planning and preparation are bad, you will see this happen too often. It costs a lot of money, so it's not what you want.

Sometimes, even when you have all the components on time, the production schedule still has to be adjusted. Anything can happen, from machine breakdown to last-minute order cancellations, from accidents to natural disasters. The whole expediting process can be for nothing because of poor planning, bad scheduling and improper execution, or just plain bad luck. In the next section, we will discuss your options in the case of such unforeseen events.

3.2.4. CHANGE THE SAIL AND DIRECT THE SHIP (CONTROL)

When you sail to a destination, the wind will not always blow from the right direction. If the winds change, you re-set the sail and off you go. Sometimes there is no wind at all. If that happens, you can paddle your ship to its destination as a last resort.

You must control the things and processes that you can control. The ones that are only within your reach, you try to influence. For the things that go beyond your reach, you prepare backup plans.

Every once in a while, something doesn't go according to plan. You have several options: stick to the plan, switch to a backup plan, or simply improvise and make up a new plan as you go along. One thing to keep in mind when changing plans: always remember the end-goal.

Why did you start up a business? One reason is to gain wealth by serving customers. Wealth will accumulate if you gain profit – revenue minus cost. You gain revenue when somebody buys your products or services. You incur costs when you buy things from suppliers, make products, and pay employees. You can control the costs and the way you do things in order to be more productive and efficient. Marketing and promotion are efforts to influence customers to buy your products or services. You try to gain more market share by influencing potential customers and keeping existing ones.

In tactical planning, when you decide to increase or decrease capacity, you can control when these changes are actually implemented. You can coordinate with the engineers and contractors that are going to execute the changes. You can shift and alter production schedules to make sure that customer orders can still be fulfilled. You can even hold some orders and push the delivery date forward. These are all things you can control. Why do you want to control them? Because there are also things that you cannot control, that may affect the ones that you can control.

For example, you may want to change the date of a capacity increase if you know the new equipment is going to be late. Another example would be if during implementation it turns out that the engineer and contractor need more time to carry out their jobs. Do you have a backup plan that took this into account? What does the backup plan tell you to do? Do you allow it to happen, or do you try to control it by having more resources to expedite the process? How will this affect your whole planning and other activities?

In a manufacturing environment, the master production schedule is a tool to synchronize the processing of the components and resources needed for production. Resources can be: capital, people, machines, energy. Given the amount of coordination involved in a production schedule, it should not be altered easily.

An MPS has several periods: the planning horizon, a fuzzy period and a frozen period. Making changes in the planning horizon is allowed. Any numbers put in there are provisional and may still change significantly. During the fuzzy period, when the moment of execution comes nearer, some of the resources are committed to process the schedule. Changes in this period are not desired, but still allowed, albeit under severe restrictions. However, when the moment of execution is near and the schedule has all the necessary details, most of the resources are committed and prepared to execute the plan. In this frozen period no changes are allowed. All efforts are to make sure that everything goes according to plan and schedule.

3.2.5. SEEING IS BELIEVING (MONITORING)

By controlling the process, you control the outcome. But if you do not watch how everything goes, how do you know if you are still in control?

In operations management, you don't take anything for granted. By monitoring operations you try to detect things that are out of order. For example, during a surgical operation there are sensors monitoring the vital signs of the patient, such as heart rate and blood pressure. While the doctors are performing the operation, these sensors constantly monitor the patient's condition. When something goes wrong, e.g. a bleeding occurs; the sensors will pick up a drop in blood pressure and set off an alarm. The doctors immediately pick up the alert and react accordingly, trying to locate the source of the bleeding. Once it's found, they will try to stop it. If the bleeding is too heavy, the doctors will ask for an additional blood transfusion to keep the patient stable and continue the operation until the whole procedure is completed.

You know you are in control if everything is within standard or expected measurements. In operations management, what do you measure? Depending on the goal of the operations, it can be as simple as the number of defects found in a production line or cost based on real activity. When you start up a company, actual cash flow compared to projected cash flow is a good measurement to tell if the business is performing according to plan. Actual sales numbers compared to forecast sales numbers can be a good indication of successful or poor marketing, sales and promotion activities. Calculating cost of goods sold and keeping a tidy bookkeeping will help you identify what is going on in the business and detect where improvements should be made.

Depending on the strategy a company has adopted, having just the right capacity is very important. Having more capacity means you have invested in more resources than you actually need. You also don't want to have less capacity than needed, because then you will not be able to satisfy the demand. How do you know if you have the right capacity? You can measure your customer backlog, inventory values and machine utilization. If you have a large customer backlog, your inventory values are at a minimum, and machine utilization is high it indicates that you don't have enough capacity. When there is no backlog and your inventory values are always high, and your machine utilization is low it means that you have too much capacity.

An unscheduled stop of a production line is a big deal. It happens only when something is out of sync, which is the result of poor coordination. Having too many line stoppages means you are not in control and that the whole planning and preparation process needs improvement.

Whenever you detect numbers deviating from standard values, it tells you to look for the cause and react accordingly. Never forget to record data and measurements regarding activities. These data may not seem important at the moment you record them, but they will start making sense after you have gathered some more and patterns start to emerge. These patterns will tell you if an activity is under control or not.

3.2.6. DO IT BETTER NEXT TIME (IMPROVEMENT)

Historical records are for improving your process in the future. Knowing is only half of the equation. What you do upon what you know completes it.

Realizing that you will repeat an event or activity several times in the future; can you do it better next time? You can identify what you didn't do well. You can reflect on the situation and conditions at that moment in time and how they came about. You can figure out a plan, either to prevent it from happening again, or to redo it. For this, you need to have data and information about the past. That is why historical records are important. With the help of these records, you can recollect and re-learn things, discuss them with others and figure out a way to improve the process. Data should be recorded all along the operations process, from when it is planned to when it is executed.

A wedding party is an example of an event that should not go wrong and will not be repeated in the future. So how can you do better if you don't have any previous experience? You can learn from similar experiences in the past. This can be based on interviewing and observing others who do have the experience, or on your own experience helping others preparing for a similar event.

For example, you can learn how to pick the venue, the catering service and the decorators. Are they reliable and can they perform well during an event? Do they provide value for money? What is the most practical and efficient way of distributing the invitations? Should there be a team to handle preparations? There are many questions that can help improve the process. First-hand experience with similar events will surely help you become better when it comes to your own wedding party.

Most people learn best and get better faster if they can either observe or experience first-hand. Do we allow a new employee to learn things by having first-hand experience? Do we risk the whole operation by letting a beginner handle the delicate process of an operation?

Can we give people such learning experiences so that they can get better in a safe environment, without risking any real damage? Can we simulate the conditions to match the real conditions of an operation? Can we introduce disruptions to show how they affect the performance of the operations? Can we fast-forward processes that require a very long time? Can we repeat them over and over again so that somebody can experience them multiple times?

All these options are available in a serious simulation game. Having multiple experiences in a short time can help people learn operations management and get better at it faster.

4. PLAYING OMG!

In this chapter we will explain how to set up, play and customize OMG! We start with a description of the physical components, followed by key concepts, gameplay (including the general rules), customization and improvement of the game.

4.1. PHYSICAL COMPONENTS

All physical components used in this game are based on printed laminated paper, colored paper and stickers.

4.1.1. TABLEMAT BOARD

The tablemat board is the representation of a production zone and is used as the base area for information cards, upgrade cards, parts as inventory or in transit..

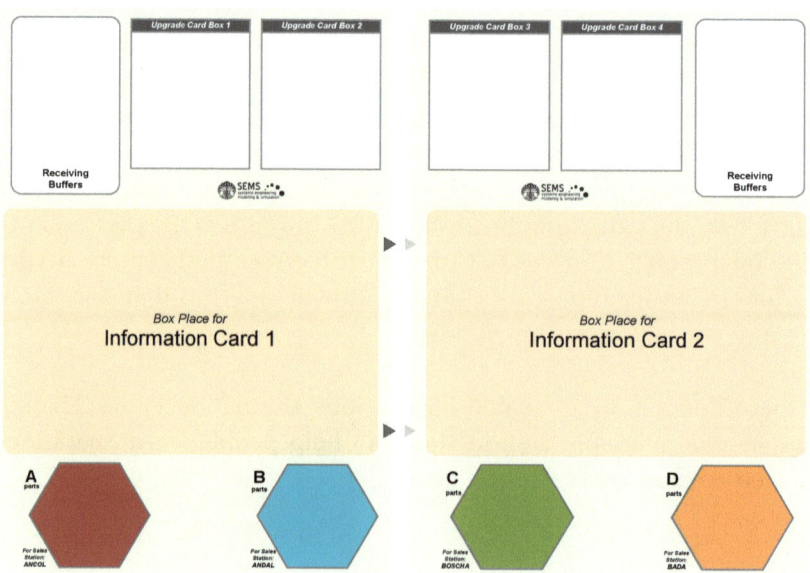

FIGURE 4-1 TABLEMAT BOARD AS THE BASE

OMG! uses two pieces of A4 sized paper as game board. This setup is customizable: it can also be three or four pieces of A4-sized paper. Figure 4-1 shows the tablemat design with two A4's

There are four major areas:
- **Information Card Area** – located at the center of the mat. The Information Cards (see Section 6.1.2) that can be placed here are the primary source of information for the players. It can hold the material flow, upgrades table, cost details and any other information the players must have to play the game.
- **Upgrade Cards Area** – located at the top of the mat, with four allocated areas. Holds all Upgrades Cards (see Section 6.1.3) purchased for a station. Each box should contain one type of Upgrade Card. Upgrade Cards of the same type with different levels should be stacked.
- **Parts-in-Process Area** – located at the bottom of the mat, with the number of areas depending on the number of parts. The colors of the boxes correspond to the colors of the parts (pieces of paper, printed stickers or paperclips – see Section 6.1.4). Additional information is added when the mat is used as the sales station and you are dealing with final products, not parts.
- **Receiving Buffer Area** – located at the corners of the mat. This area accommodates material inflow from the left or right, depending on the table configuration. This is an optional area, although it is recommended as a visual sign to keep track of incoming buffer stock.

4.1.2. INFORMATION CARDS

The Information Cards are a game component in the form of two pieces of A5-sized paper. The cards contain basic information about each station's role and standard conditions.

There are two areas available in which you can put any information (Information Card 1 and Information Card 2).

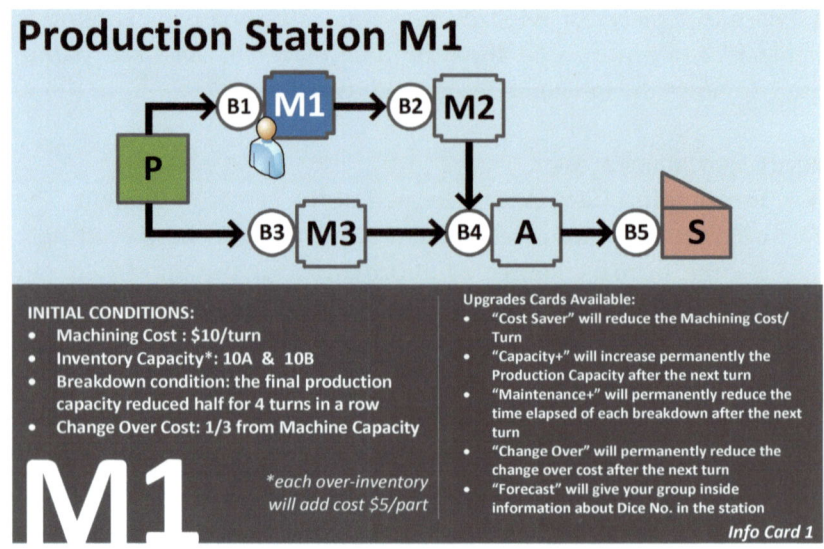

FIGURE 4-2 INFORMATION CARD 1 FOR PRODUCTION STATION

For example, Information Card 1 for Production Station 1, as shown in Figure 4-2, shows the operation chain and the place where M1 is positioned in it. It also contains information about the initial conditions, machining costs, inventory capacity, breakdown info, changeover costs, and if upgrade cards are available for that station.

Information Card 2 for Production Station 1, as shown in Figure 4-3, contains the rules of the game, the dice translation table, upgrade card costs and impacts.

Production Station M1

- Your station are producing parts A and parts B, then send them to the next station's (M2) receiving buffers as many as the dice number (Table 1).
- For every turn you can send ONLY 1 type of items (either A or B).
- If the item you sent to the next station differs from the one you sent before, you will get a change over cost (represented by reduction of production capacity at that turn)
- You can only send parts from your current inventory, not from the receiving buffers. After you sent the parts, you can move parts from the receiving buffer into your inventory (based on type) and empty the receiving buffers.
- Your Station could buy available upgrade cards listed in Table 2

M1

Table 1 Dice Translations

Dice #	Machine Capacity
1	1
2	2
3	3
4	4
5	5
6	6

Table 2 Upgrade-Cards Value Table

Level	Cost Saver		Capacity+		Maintenance+	
	$	price	pcs	price	pcs	price
0	10	-	Table	-	Table	-
1	7	25	+1	1000	+1	1000
2	4	50	+2	2000	+2	2000
3	1	75	+3	3000	+3	3000

Level	Change Over+		Forecast	
	cap	price	info	price
0	2/3	0	0	0
1	3/4	1000	Average	1000
2	Table	2000	Histogram	2000
3			Trend	3000

- Fraction always round up (0.05 = 1)

Info Card 2

FIGURE 4-3 INFORMATION CARD 2 FOR PRODUCTION STATION

The Assembly Station, which differs in its function from the other production stations because here the parts are assembled, has additional information on Information Card 2, such as an assembly identification table, as shown in Figure 4-4.

The Purchasing Station is simpler, with only the game rules and the dice translation table on Information Card 2, as shown in Figure 4-5.

43

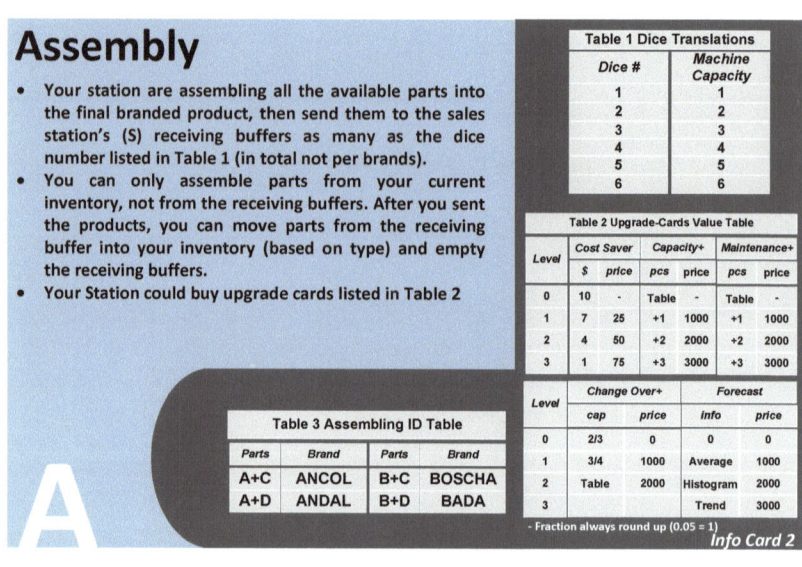

FIGURE 4-4 INFORMATION CARD 2 FOR ASSEMBLY STATION

Purchasing

- Your Station is responsible for purchasing parts of A, B, C, and D, then send them to the corresponding station (to the receiving buffers) for each type:
    Parts A or B sent to Buffer M1,
    Parts C or D sent to Buffer M3.

- The number sent is based on Table 1 for next station receiving buffer. You have 2 downstream lines (M1 and M3), and for each line you can only send 1 type of parts for each corresponding line receiving buffer. *(eg. If you have Dice 1, then you can send 3A to M1 and send 3D to M3, or 3B to M1 and 3C to M3)*

Table 1 Dice Translations

Dice #	Real #
1	3
2	3
3	3
4	4
5	4
6	4

Info Card 2

FIGURE 4-5 INFORMATION CARD 2 FOR PURCHASING STATION

The Sales Station focuses on backlog cost, price and demand, as shown in Figure 4-6 and Figure 4-7.

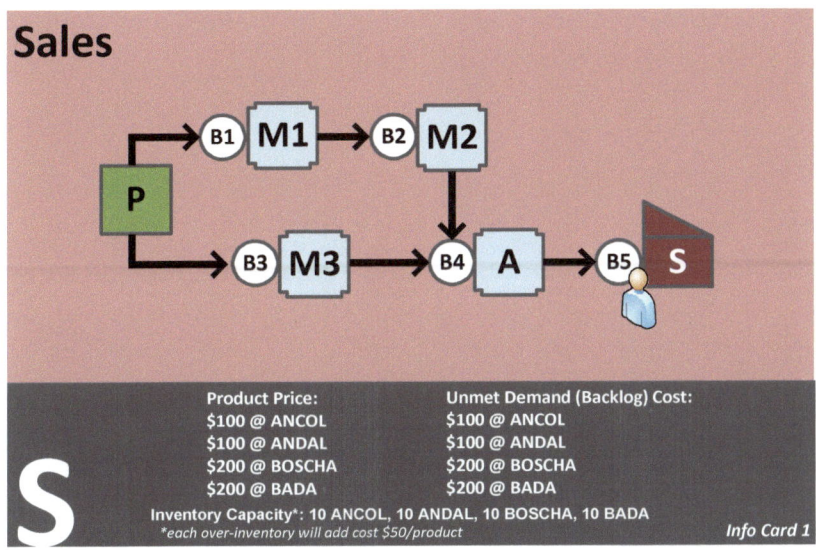

FIGURE 4-6 INFORMATION CARD 1 FOR SALES STATION

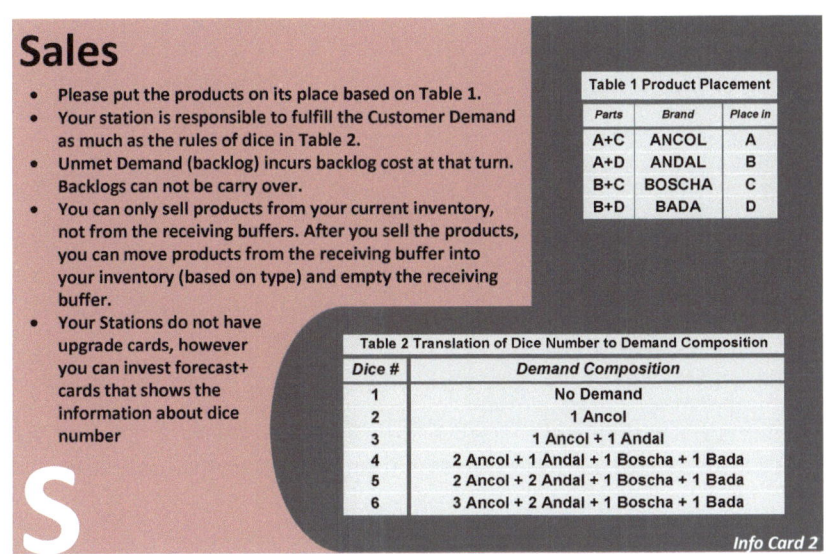

FIGURE 4-7 INFORMATION CARD 2 FOR SALES STATION

45

The complete designs for all information cards are shown in Figure 4-8.

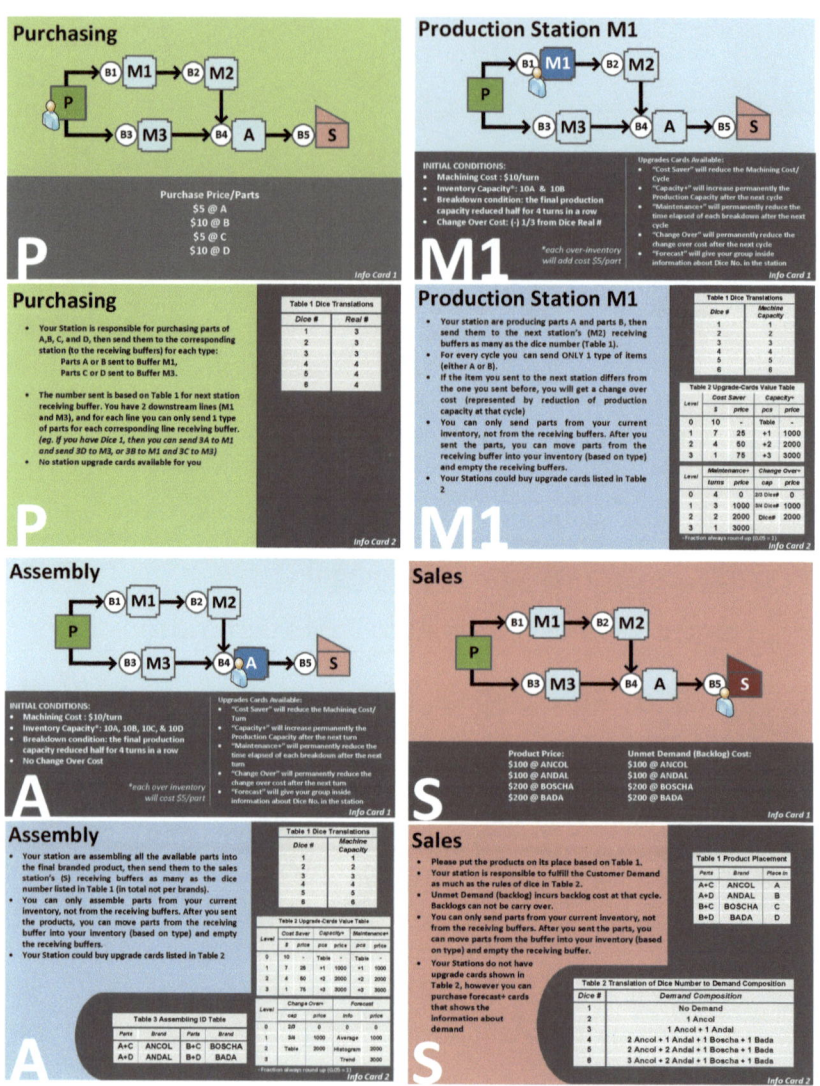

FIGURE 4-8 INFORMATION CARDS FOR ALL FOUR STATION TYPES

The information cards are dynamic, depending on how you have customized the game. Therefore, you can put as much information on them as is relevant to the game.

4.1.3. UPGRADE CARDS

The Upgrade Cards represent strategies that teams can choose to improve their operations. They are in the form of a small card (5.7 x 7 cm) as shown in Figure 4-9.

FIGURE 4-9 UPGRADE CARD DESIGN

At the top, the Upgrade Card has a title and description of what it will do. It has a simple icon in the middle for easy identification, next to the price of the upgrade. At the bottom there is the card level identification.

The various Upgrade Cards are shown in Figure 4-8.

47

FIGURE 4-10 UPGRADE CARDS TO IMPROVE OPERATIONS

4.1.4. MATERIALS

Materials are available for representing parts to be used in production, for example: in the form of pieces of colored paper and printed plain stickers (shown in Figure 4-11).

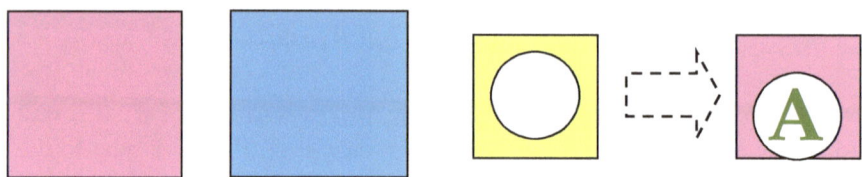

FIGURE 4-11 PAPER AND STICKERS AS PARTS REPRESENTATION

4.1.5. COMBINING THE ELEMENTS ON THE TABLEMAT

Figure 4-12, Figure 4-13 and Figure 4-14 illustrate how all components of the game are used during gameplay.

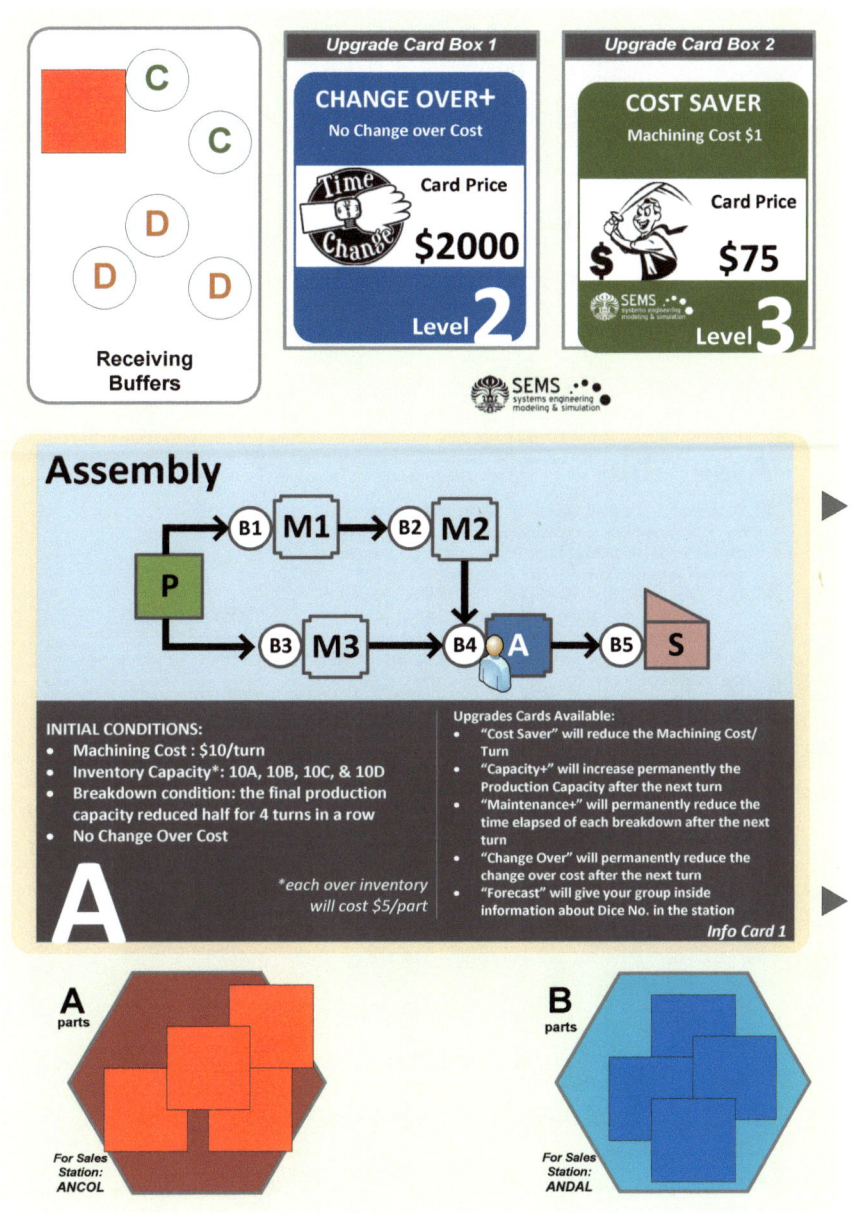

FIGURE 4-12 EXAMPLE OF TABLEMAT PAGE 1 IN GAMEPLAY

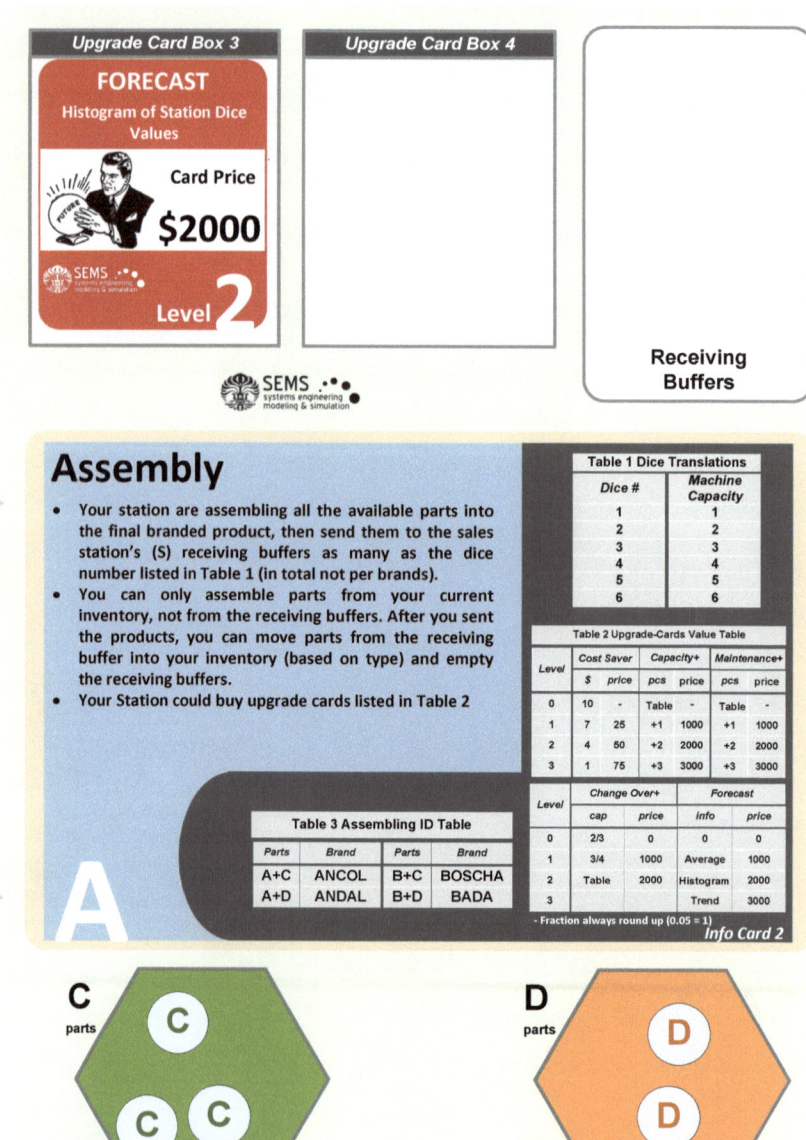

FIGURE 4-13 EXAMPLE OF TABLEMAT PAGE 2 IN GAMEPLAY

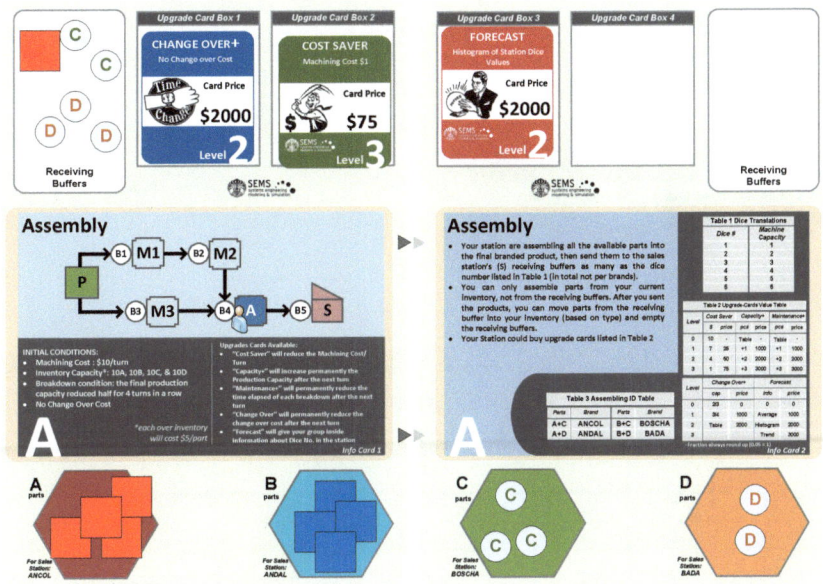

FIGURE 4-14 BOTH PAGES OF TABLEMAT SIDE BY SIDE IN GAMEPLAY

4.1.6. Record Forms

Record Forms are A4 papers on which the conditions and costs for each station are registered. These help players monitor their performance and choose the best strategy for their team. It will also help the debriefing process, where information from the forms will help players understand the impact of their decisions during the game.

Each station has its own specific record forms, from purchasing station form, production station form, assembly station form and sales station form.

We will show you the examples for all forms.

- **Purchasing Station Form**

The Purchasing Station Form records the amount of raw materials purchased along with the cost of each respective part.

Turn	Dice No.	Parts Purchase					Cost				Cost
		A	B	C	D		A	B	C	D	
1											
2											
1											
2											
3											
4											
5											
6											
7											
8											
9											
10											
11											
12											
13											
14											
15											
16											
17											
18											
19											
20											

FIGURE 4-15 PURCHASING STATION FORM EXAMPLES

- **Production Station Form**

The Production Station Form records the variability of production capacity, its impact on the inventory, and the impact of upgrades on the station.

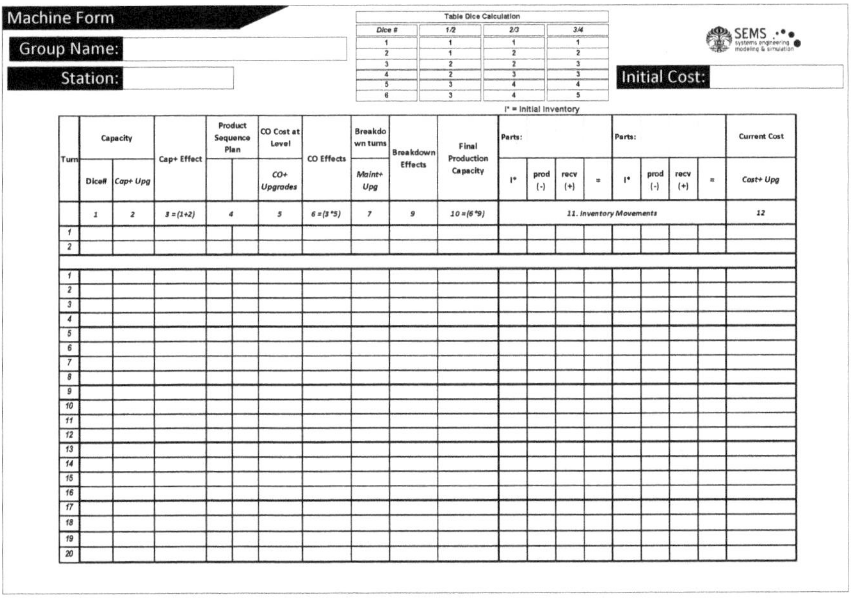

Machine Form

Group Name:

Station:

Initial Cost:

Table Dice Calculation			
Dice #	1/2	2/3	3/4
1	1	1	1
2	1	2	2
3	2	2	3
4	2	3	3
5	3	4	4
6	3	4	5

I* = Initial Inventory

Turn	Capacity		Cap+ Effect	Product Sequence Plan		CO Cost at Level	CO Effects	Breakdown turns	Breakdown Effects	Final Production Capacity	Parts:				Parts:				Current Cost
	Dice#	Cap+ Upg				CO+ Upgrades		Maint+ Upg			I*	prod (-)	recv (+)	=	I*	prod (-)	recv (+)	=	Cost+ Upg
	1	2	3 =(1+2)	4		5	6 =(3*5)	7	9	10 =(6*9)	11. Inventory Movements								12
1																			
2																			
1																			
2																			
3																			
4																			
5																			
6																			
7																			
8																			
9																			
10																			
11																			
12																			
13																			
14																			
15																			
16																			
17																			
18																			
19																			
20																			

FIGURE 4-16 PRODUCTION STATION FORM EXAMPLE

- **Assembly Station Form**

The Production Station Form records the variability of production capacity, its impact on the inventory, and the impact of upgrades on the station.

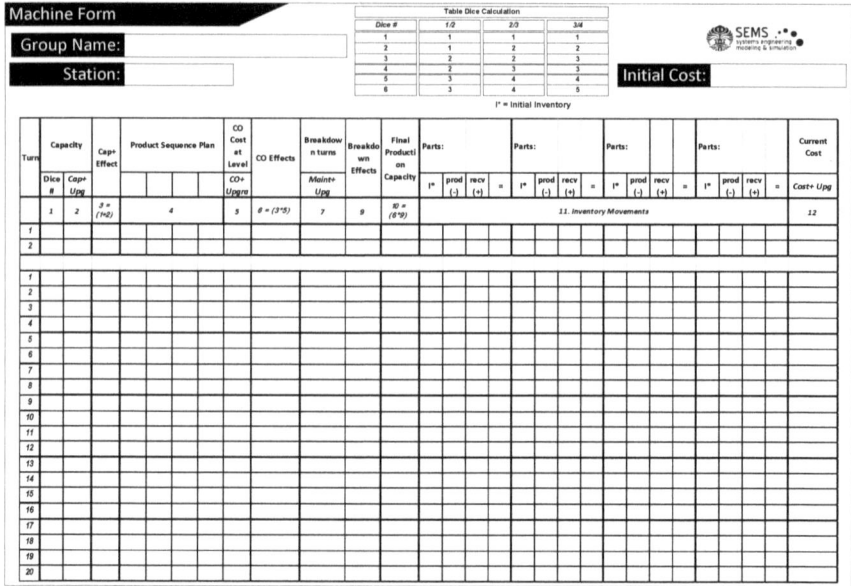

FIGURE 4-17 ASSEMBLY STATION FORM EXAMPLES

- **Sales Station Form**

The Sales Station Form records the production results to the customer, as well as demand fulfillment and the end inventory.

Turn	Initial Inventory (1)				Dice Demand	Delivered (2)				$$$ Sold (a)	Backlog Cost (b)	Total Income (a)-(b)	Input from previous station (3)				End Inventory (1-2+3)			
	AC	AD	BC	BD		AC	AD	BC	BD				AC	AD	BC	BD	AC	AD	BC	BD
1																				
2																				
1																				
2																				
3																				
4																				
5																				
6																				
7																				
8																				
9																				
10																				
11																				
12																				
13																				
14																				
15																				
16																				
17																				
18																				
19																				
20																				

FIGURE 4-18 SALES STATION FORM EXAMPLES

4.1.7. DICE THROWS OR RANDOM NUMBER SPREADSHEET

The game shows the complexity of operations decisions, one of which is variability that occurs at all stations during production. Variability can be emulated using a dice, except in the case of multi-team play. For this we have developed a dataset using a randomizer spreadsheet, so that every team experiences the same variability.

4.2. Key Concepts

The key concepts used in OMG! are: assignable costs, upgrading and variability.

4.2.1. Assignable Costs

In every production activity money is a critical factor that is usually measured against the production activity result.

The production activity itself incurs general costs:
- **Material costs**: expenses from buying raw material from suppliers or vendors. The material costs are calculated by multiplying the material quantity with the unit price.
- **Machining costs**: expenses from running the machine when it works. The machining costs are incurred at each production station. When a station has produced a part and transfers it to the next station, it will cost the team a fixed amount.
- **Over-inventory costs**: expenses from each part or product that is over the inventory capacity limit.
- **Backlog costs**: expenses from unfulfilled orders at the Sales Station.

All costs will be calculated per turn and recorded on the Company Overview Form.

Additional costs that can be included are:
- **Transportation costs**: expenses from getting raw materials from another company that occur if third party logistics are used. For example, each additional raw material delivery gives you $5 extra transportation costs.
- **WIP costs**: work-in-process parts that stay too long in one station can cause reduction of parts quality. For example, parts that have stayed at a station for two turns cannot be used anymore, which will give you a penalty in the form of WIP costs or loss of those parts.

4.2.2. UPGRADING

Money on hand can be used to invest in improved performance of the production system in the form of Upgrade Cards. Every Upgrade Card, except for the Forecast Card, is valid for one station only. You can also modify this to let any Upgrade Card affect one or more stations, or the whole operations chain.

The basic Upgrade Cards are divided into five categories:
- **Cost saver**: This card reduces the machining cost in the production and assembly stations for each production cycle. The upgrades are final and not time-limited. You can define how much the reduction costs are in absolute costs or relative to the original costs. The higher the level of the upgrade card, the lower the machining costs.
- **Capacity+**: This card increases the maximum production capacity in the production and assembly stations for each production cycle. The upgrades are final and not time-limited. You can define how many parts can be produced at different levels in absolute parts or relative to the original amount of parts. The higher the level of the upgrade card, the higher the increase of production capacity.
- **Maintenance+**: This card reduces the amount of turns that are impacted by a breakdown. The upgrades are final and not time-limited. You can define how many turns are impacted when a breakdown occurs. The higher the level of the Upgrade Card, the lower the number of turns that are impacted by breakdown.
- **Change Over+**: This card reduces the changeover penalty. Level 1 reduces the penalty to ¾ of the dice number, whereas level 2 will get rid of the changeover penalty altogether.
- **Forecast***: This card contains the information corresponding to the dice number for one station for the whole game. Level 1 gives the average amount, level 2 gives a histogram for the dice number (1-6) frequency; level 3 gives the dice number trend for the whole game.

4.2.3. VARIABILITY

There are breakdown scenarios that should be addressed by the players. For example, OMG! can simulate the conditions of planned maintenance breakdowns. Also, in every production process variability inevitably occurs in the form of unforeseen circumstances, from machine breakdown to accidents and from temporary personnel shortage to a surge in demand.

In case of single-team play, a physical dice is used at each station to represent the randomness distribution that exists in any production process and influences machine capacity.

TABLE 4-1 COMBINATION 1 OF DICE TRANSLATION TABLE

Dice Number	Machine Capacity
1	1
2	2
3	3
4	4
5	5
6	6

In order to reduce the span of randomness you can translate the dice numbers as in Table 6-2.

TABLE 4-2 COMBINATION 2 OF DICE TRANSLATION TABLE

Dice Number	Machine Capacity
1 & 2	2
3 & 4	4
5 & 6	5
5	5

You can also decide not to use the dice, which is limited to the numbers 1 to 6. You can assign a probability number to represent process variability.

4.3. GAMEPLAY

4.3.1. GAME STAGES

OMG! involves three stages of activities:
1. Briefing – before the game
   A brief introduction to familiarize the players with the game. This consists of the rules of the game and how to play. Game goals must be explained to enhance knowledge transfer through the game.

2. Playing – during the game
   During gameplay, the players have to make their own decisions regarding the simulated production activities. Their speed in making decisions will also impact the dynamics of the game flow. At this stage, the players' interaction within their own team is the main activity. The facilitator reminds them about the game steps and provides the scenarios. All the things that the players are expected to learn are implicitly experienced within the game's activities. The players experience that playing the game is a learning process.

3. Debriefing – after the game
   During the debriefing, all participants and the facilitators discuss and share their experiences, and conclude the learning process they went through.
   At this stage, the facilitator should guide the discussion, covering:
   - The learning points the players experienced
   - The impacts of their planning decisions at each station on team performance
   - The impacts of their investments
   - The single most effective investment
   - Their ability to handle variability and breakdown

The debriefing stage will help in evaluating whether the goals of the game have been achieved or not. All feedback from the game participants is very precious to make the game more suitable for achieving the targeted learning points.

4.3.2. Rules

The general rules of the game are:
- One person is responsible for one station. However, it is also possible that two or more people are responsible for one station.
- In order to have a smoother production flow, each player or team is given a certain amount of initial inventory at the start of the game. The initial inventory can be final products or parts. This is the first tactical decision the players should make.
- We use cycles and turns. One cycle indicates the time needed to fulfill the production plan. The production plan consists of what to buy, and what to produce at what stations. One cycle contains several turns (e.g. 1 cycle = 4 turns). During turns, the players cannot change their production plan. At the end of a cycle, the players have to plan the parts and products to be produced in the next cycle. Remember there are changeover costs and other costs to consider.
- Investments on Upgrade Cards are available when a player has sufficient money on hand. Upgrades can only be purchased between cycles.

The most successful player or team in this game is the one who can fulfill customer demand and has the most money on hand at the end of the game. Each fulfilled order will add to your profit, while every unfulfilled order will give you a penalty in the form of backlog cost, which reduces your profit. Every production activity and material purchase involves expenditures. All profits, costs and investments will be totaled at the end of the game.

4.3.3. Game Steps

1. Decide product plan sequence
2. Roll the dice
3. Translate the dice number and send parts/products to the next station's receiving buffer
4. Record your current conditions in your station record form
5. Report your station cost/income/upgrades to the Purchasing Station

6. Move the parts/products from the receiving buffers to the inventory area
7. Report to the facilitator for upgrade investments
8. Restart for the Next Periode

4.4. Customizing OMG!

OMG! can be customized to meet the participants' needs. In this section we will describe some of customization that you can do in OMG!

4.4.1. Customization of Product Variation

The game can be played in a very simple way with design A, or in a more complex way with design C. The impact of customization as shown in Table 4-1 will be explained below.

TABLE 4-3 POSSIBLE COMBINATIONS OF OMG! PRODUCTION

OMG! Design	No. of Stations	No. of Parts	No. of Final Products
A	5	3	2
B	6	4	4
C	8	5	6

Modification of the number of stations in OMG! will not change the production flow or the game configuration layout.

4.4.2. Design A – 5 stations, 3 parts and 2 products

This configuration is the simplest configuration of OMG!, with four primary roles for the stations:
- P = Purchasing Station
- M1 = Production Station M1
- M2 = Production Station M2
- A = Assembly Station
- S = Sales Station

61

The suggested material flow is illustrated in Figure 4-19

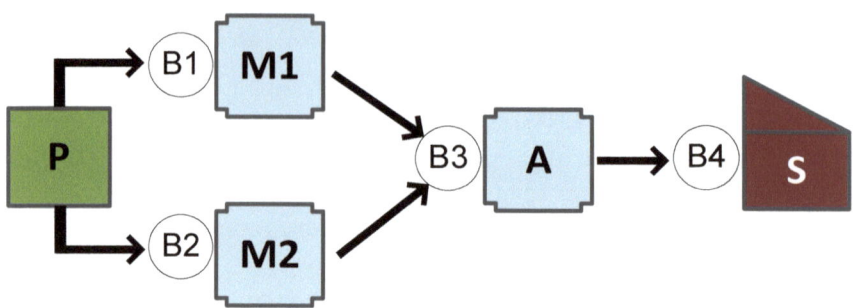

FIGURE 4-19 FLOW OF MATERIALS FOR 5 STATIONS

This material flow design with five players can be configured as shown in Figure 5-3.

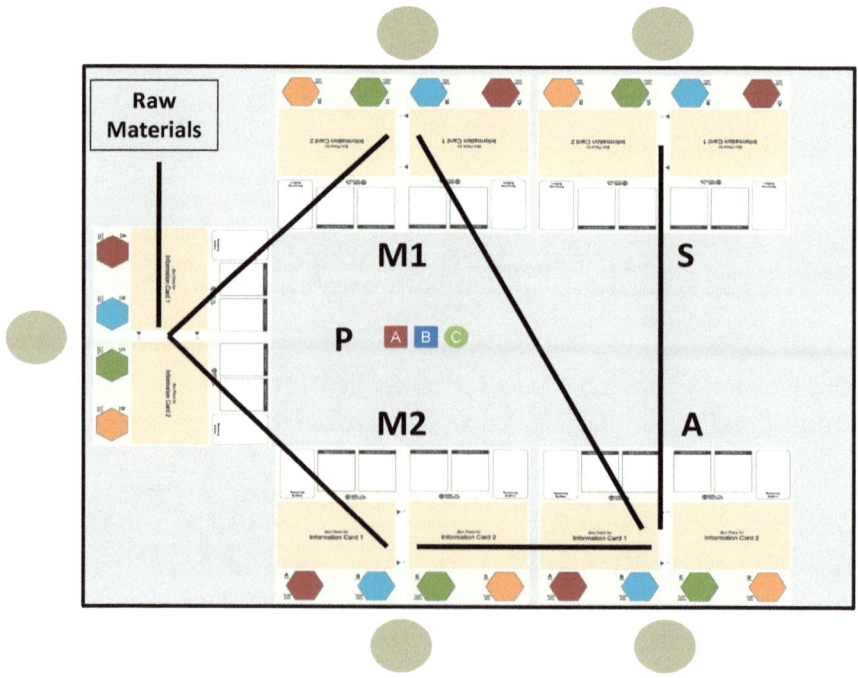

FIGURE 4-20 RECOMMENDED TABLE LAYOUT FOR 5 STATIONS

Product variation is limited to two products using combinations of two different parts (listed in Table 4-4).

TABLE 4-4 2 PRODUCTS AND 2 PARTS COMBINATIONS

Product Names	Parts Combinations
Ambon	AB
Acacia	AC

4.4.3. DESIGN B – 6 STATIONS, 4 PARTS AND 4 PRODUCTS

This configuration is the recommended configuration for playing OMG! Chapter 7 provides an example of playing with this configuration. It consists of one Purchasing Station (P), three Production Stations (M1, M2, and M3), with one Assembly Station (A), and one Sales Station (S).

The material flow is shown in Figure 4-21.

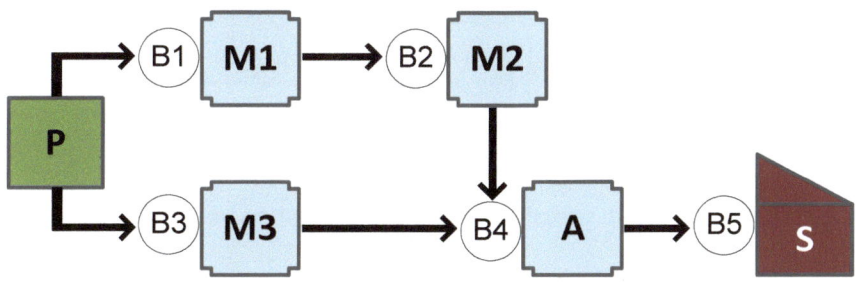

FIGURE 4-21 FLOW OF MATERIALS FOR 6 STATIONS

Product variation is limited to four products using combinations of four different parts (listed in Table 4-5).

TABLE 4-5 4 PRODUCTS AND 3 PARTS COMBINATION

Product Names	Parts Combinations
Ancol	AB
Andal	AC
Boscha	BC
Badai	BD

The table layout for this configuration is shown in Figure 4-22.

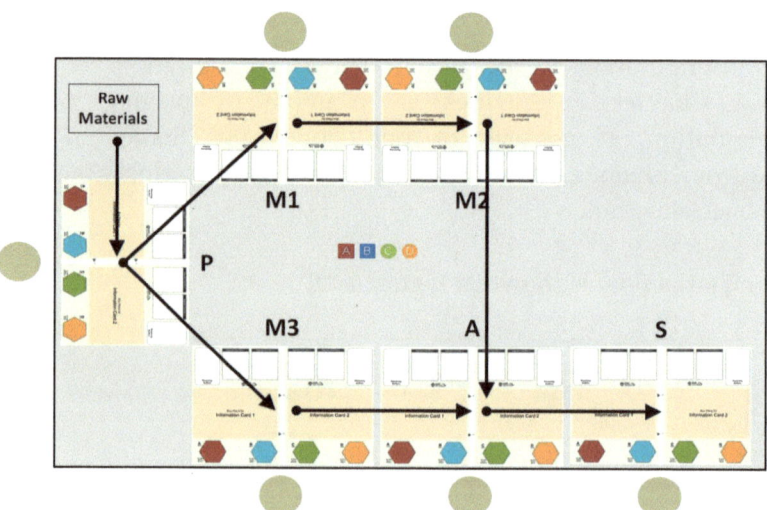

FIGURE 4-22 RECOMMENDED TABLE LAYOUT FOR 6 STATIONS

4.4.4. DESIGN C – 8 STATION, 5 PARTS AND 6 PRODUCTS

This configuration consists of one Purchasing Station (P), four Production Stations (M1, M2, M3, and M4), two Assembly Stations (A1 and A2), and one Sales Station (S). The material flow is shown in Figure 4-23.

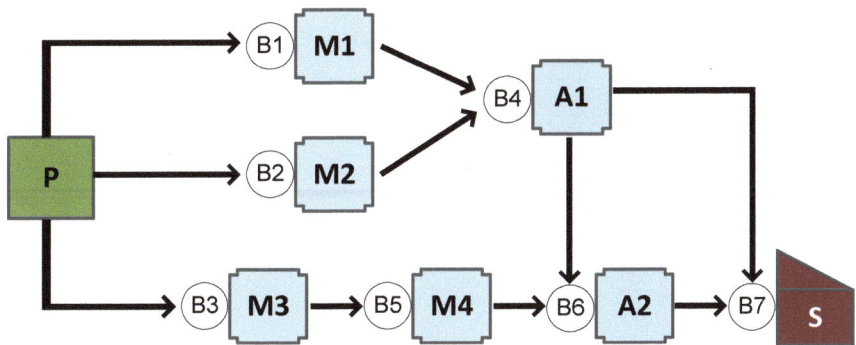

FIGURE 4-23 FLOW OF MATERIALS FOR 8 STATIONS

The table layout for this configuration is shown in Figure 4-24.

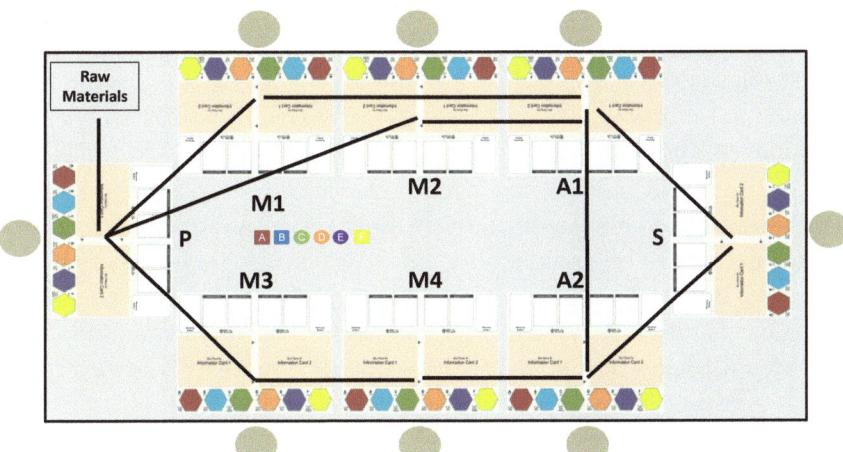

FIGURE 4-24 RECOMMENDED TABLE LAYOUT FOR 8 STATIONS

Product variation is limited to six products using combinations of five different parts (listed in Table 4-6).

TABLE 4-6 6 PRODUCTS AND 5 PARTS COMBINATION

Products Name	Parts Combination
Ancol	AB
Andal	AC
Andal Premium	ACE
Bosscha	BC
Badai	BD
Boscha Premium	BCE

4.4.5. CUSTOMIZATION OF UPGRADES

When thinking about new upgrade cards, the starting point are the variables and constraints in the game. These can affect costs, penalties, time, demand and revenue.

You need to use different prices of investment as a way to reflect proportionality in the real world. It would be illogical, for example, if cost saver upgrades are more expensive than capacity upgrades, since in the real world a capacity upgrade means buying more costly machines or upgrades.

You also need to balance income capability and investment. You don't want the teams to be unable to invest at all because their income is not sufficient during the game. Setting the right price for worst-case scenarios is crucial. Worst-case scenarios are for example when breakdown occurs frequently or when the dice number is lowest more than 75% of the time.

Some additional upgrade cards that you could include are:
- **Discount Card**: This card reduces the cost of material purchases. Every level of upgrade corresponds to a different discount value.
- **Advertising Card**: This card allows you to sell more than the actual demand, which helps you make more money. It affects

either certain final products or all final products. The higher the upgrade level, the more additional final products you are allowed to sell.

TABLE 4-7 UPGRADE CARDS INVESTMENT EXAMPLE WITH 3 LEVELS

Cards Type	Level 1	Level 2	Level 3
Cost Saver	$25	$50	$70
Capacity+	$1,000	$2,000	$3,000
Maintenance+	$1,000	$2,000	$3,000
Change over+	$1,000	$2,000	$3,000
Forecast	$1,000	$2,000	$3,000

4.4.6. CUSTOMIZATION OF VARIABILITY

It's possible to introduce variability that affects gameplay, such as the introduction of:
- **Breakdown**: based on the demand pattern used in the game you can introduce breakdown conditions randomly or as planned maintenance. Random breakdowns could create a more level playing field, especially if one team looks to dominate. Predictable maintenance at scheduled intervals would introduce a more complex production schedule for the players to develop. Breakdowns could lead to capacity reduction or additional costs at certain intervals.
- **Local or international holidays:** In some cases, holidays can mean a combination of demand surge and limited production capacity. Introducing this condition increases the complexity of production scheduling.
- **Demand**: The variability of demand can be modified. The dice can be used to represent the randomness distribution. You can also create a translation of how these values impact each station or cause demand variation.

An example of demand variation is listed in Table 4-8. You can reduce the variation of the dice by combining its values as in Table 4-9.

TABLE 4-8 DEMAND TRANSLATION TABLE EXAMPLE

Dice number	Demand Composition
1	0
2	1 ANCOL
3	1 ANCOL, 1 ANDAL
4	2 ANCOL, 1 ANDAL, 1 BOSSCHA, 1 BADA
5	2 ANCOL, 2 ANDAL, 1 BOSSCHA, 1 BADA
6	3 ANCOL, 2 ANDAL, 1 BOSSCHA, 1 BADA

TABLE 4-9 MODIFIED DEMAND TRANSLATION TABLE EXAMPLE

Dice number	Demand Composition
1	0
2 & 3	1 ANCOL, 1 ANDAL
4	2 ANCOL, 1 ANDAL, 1 BOSSCHA, 1 BADA
5 & 6	3 ANCOL, 2 ANDAL, 1 BOSSCHA, 1 BADA

4.4.7. Customization of Performance Variables

The basic performance variables of OMG! focus on profit, which is defined as revenue minus cost and balanced product supply (no products left behind).

Additional performance variables can be introduced, such as:
- **Delay time**: defined as the time (turns or cycles) used to produce products from purchasing to market. You can adjust the production delay time from the Purchasing Station to the Production Stations or within the Production or Assembly Stations and from Production or Assembly Stations to the Sales Station.
- **Inventory turnover**: the inventory period time (turns or cycles). Game materials in the inventory area for X turns or X cycles are counted as performance variable.
- **Assembled product quality**: product quality can be introduced by having criteria for accepted products. This would depend on the design of the materials for parts. If you are using paper and round stickers for example, you could make a small circle

on the paper and judge if the stickers are aligned with that circle. It is also possible to judge a group of every five final products instead of each product.

4.5. Recommended Steps in Customizing OMG!

There are five steps that we recommend you go through when customizing OMG! The process should be iterative, which means that it is possible to return to the previous step if during one step you find that it is impractical or impossible to continue.

4.5.1. Define Game Objectives

Defining objectives at the beginning of the customization process is important to provide a broad guideline, since there is an almost unlimited amount of configurations you can make of OMG!

Some questions you may need to answer are:
- Are you using the game for competition only or for learning purposes as well?
- How many learning points do you aim to transfer to the students?
- How much time do you have for gameplay and is it continuous or spanning different days?
- Are you going to have multiple sessions during which the team can test different strategies based on their previous experiences?
- Are there limitations in the facilities or materials you need to consider?

4.5.2. Translate Objectives Into Customization Variables, Constraints And Upgrades

After the game objectives are set, you can define the variables and constraints that will be available in the game.

Some questions you may need to answer are:
- How many variables and constraints do you need to achieve the game objectives?
- What variable priorities do you need the participants to have based on the game objectives?

- Is there a reference decision sequence that participants must use to gain a reasonable profit?
- At which periods do you allow the participants to have enough money for upgrades (by modifying the price and cost of products, parts and upgrade cards).
- Do some upgrades have more impact than other upgrades, and what are the ratios between them?
- What scenarios do you need to add so the game will be more challenging?
- Is there a real-case scenario that reflects the necessary learning goals?
- Is there a demand pattern that you need the participants to understand?

4.5.3. Re-Design Components (Tablemat, Parts, Upgrade Cards, Forms)

The tablemat and all the forms and upgrade cards can be redesigned to accommodate changes in the OMG! configuration.

You can redesign:
- **Parts**: decide how you will represent parts in the game. They should be big enough to distinguish, but not space-consuming. They should also be light (too heavy could cause accidents).
- **Tablemat**: must accommodate the size of the parts and product variations.
- **Forms**: you must put yourself in the participants' shoes and think about the sequence of the information they must put on the forms. The forms should make it easy for the participants to monitor any upgrades they have bought.
- **Pre-briefing** and **briefing materials**: it may be necessary to develop a manual for the participants to read prior to the game. In it you can put theoretical references necessary to understand the game components.
- Spreadsheet of **demand patterns**: you can use any spreadsheet application to generate random numbers for the game.
- Pre-game, in-game or post-game **test and questionnaire**: if necessary, you can conduct a pre-game questionnaire to make sure that the participants did their homework and read all the

materials provided. During the game, you can pause the game and conduct a test, which is also a way of giving hints to the participants on what they must consider. A post-game test could help make sure that the participants keep their mind in a reflective mode for learning, because they know that they will be tested after the game.

4.6. PROTOTYPING AND FINALIZING THE GAME

When the components are ready, create a prototyping group to test your set-up. Create a discussion to gain feedback about the game components, playing sequence and game variables. It is also advisable to develop prototypes with multiple groups if you are going to have multiple groups playing OMG!

This is a step where you may have to return to previous steps iteratively. For example, you may need to revise your game objectives, configuration and design of components.

4.7. EVALUATION AND CONTINUOUS IMPROVEMENT

There can always be things that we may have overlooked in the design and prototyping steps, which means there is room for improvement. You can develop a special questionnaire to get feedback from the participants. However, you should be careful that sometimes participants complain about difficulties that are actually part of the challenges of the game.

5. OMG! EXAMPLE WITH 6 STATIONS

In this chapter, we will explain how to play OMG! with six stations and four products using combinations of four different parts. In this case, the game is played by six people in one team (single-team play).

Each station is managed by one player. The products are: Ancol, Andal, Bosscha, and Bada. (As you may have guessed, the names of the products are based on a combination of two parts named A, B, C, and D.)

The material flow is shown in Figure 5-1. You can play this set-up on a regular table, a square table, or a round table. When deciding the layout, please remember to consider that the players must be able to move parts from one station to the next without too much effort.

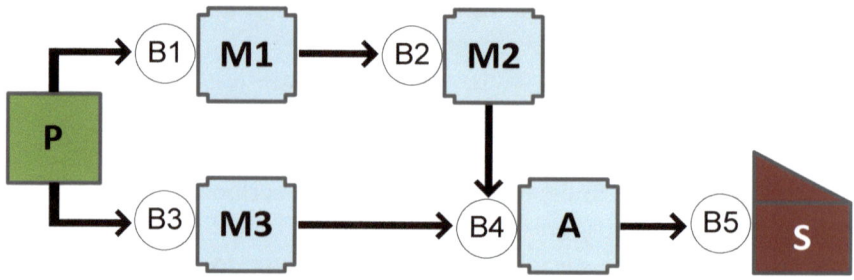

FIGURE 5-1 FLOW OF MATERIALS FOR 6 STATIONS EXAMPLE

Raw materials for branded products A, B, C, and D are sent from the Purchasing Station (P) to Production Stations M1 and M3. Station M1 then sends parts A and B to Station M2. Stations M2 and M3 send the parts to the Assembly Station (A). The Assembly Station then assembles the final products (Ancol, Andal, Bosscha, and Bada) depending on demand. The Assembly Station sends the finished products to the Sales Station (S).

In a single-team environment, each station uses the dice for setting the maximum capacity to send parts or products. The dice number for each station is translated according to the dice translation table on the Information Card.

5.1. SUGGESTED TABLE LAYOUT

When using a square table, the six persons should be seated within arm's length from each other. This is not an absolute requirement, as long as the flow of material is visible and each player's view of the stations before and after his/her own station isn't blocked. Players should also be able to look at the current inventory of their production chain. Therefore the layout suggested in Figure 5-2 and Figure 5-3 is the recommended layout.

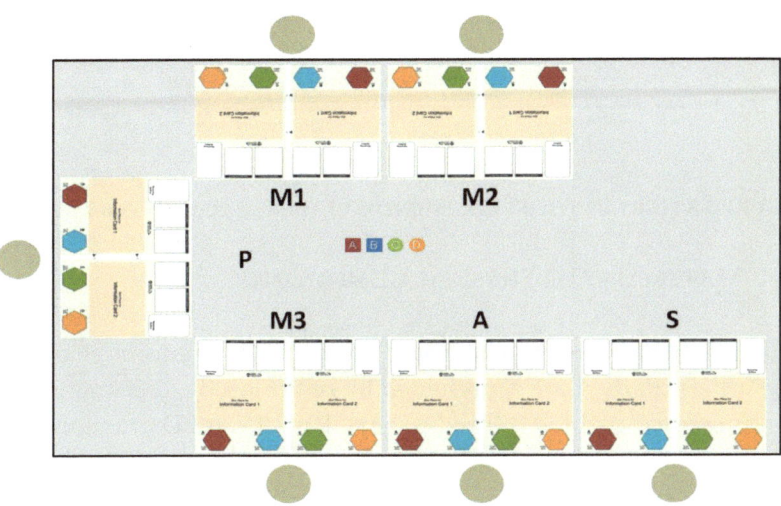

FIGURE 5-2 RECOMMENDED TABLE LAYOUT FOR 6 STATIONS

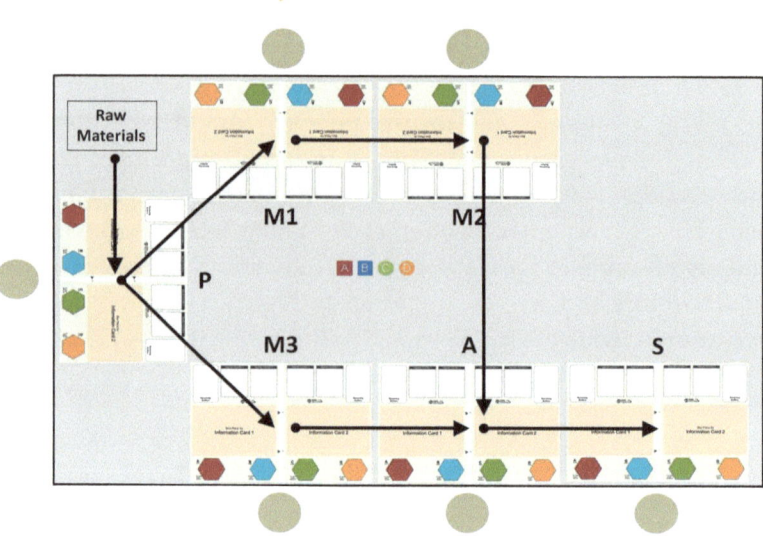

FIGURE 5-3 FLOW IN THE RECOMMENDED TABLE LAYOUT FOR 6 STATIONS

5.2. MATERIALS FOR PHYSICAL COMPONENTS

Using off-the-shelf materials, OMG! combines colored pieces of paper and stickers. Parts A are represented by red colored papers and parts B are represented by blue colored papers. Parts C and D are represented by plain circular sticker labels identified by the letters C and D. You can use pre-printed stickers, or just write the letter C or D prior to delivering them to their respective stations.

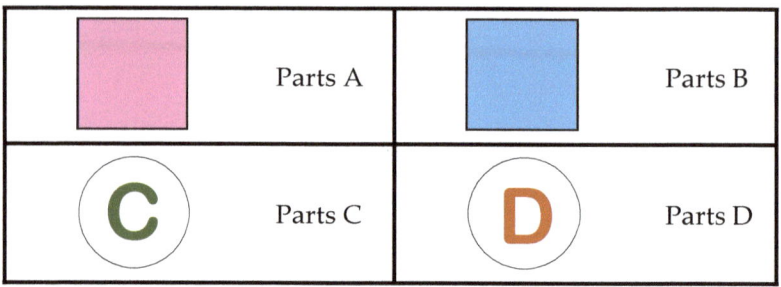

FIGURE 5-4 ILLUSTRATION OF 4 PARTS

5.3. STATION DESCRIPTIONS

5.3.1. PURCHASING STATION

The player at the Purchasing Station is responsible for supplying parts A, B, C, and D to Production Stations M1 and M3. Parts A and B are sent to Station M1 and parts C and D are sent to Station M3. The Purchasing Station has to supply both Station M1 and Station M3 at each turn. At each turn, it can only send two types of parts. The number of parts sent to the next station is equal to the dice number, as listed in Table 5-1.

TABLE 5-1 DICE TRANSLATION TABLE (PURCHASING STATION)

Dice Number	Real Number
1	3
2	3
3	3
4	4
5	4
6	4

For example, if a station gets dice number 1, it can sent a combination of three parts A to M1 and three parts C to M3, or a combination of three parts B to M1 and three parts D to M3.

5.3.2. PRODUCTION STATION (M1, M2, AND M3)

The players at the production stations are responsible for processing the parts from their current inventory to the next station's receiving buffer.

Each production activity costs $10 (machining cost) per turn. You can only send one type of part in your inventory at each turn. When you want to change to a different type of part at the next turn, a changeover cost will be charged. This will reduce your maximum production capacity to 2/3 for that turn.

During the game, breakdown conditions may occur in the production stations at random. They will reduce your maximum capacity to ½ for four turns in a row, starting with the current turn.

Inventory capacity for each station is ten pieces of each type of part, whereas the maximum amount of processed parts to be transferred to the next station is determined by the dice number (as listed in Table 5-2).

TABLE 5-2 DICE TRANSLATION TABLE (PRODUCTION STATIONS M1, M2, M3)

Dice Number	Machine Capacity
1	1
2	2
3	3
4	4
5	5
6	6

5.3.3. ASSEMBLY STATION

The Assembly Station assembles the parts from the current inventory and sends them as final products to the Sales Station's receiving buffer. It has the same properties as the Production Stations, which include:

- Each production activity costs $10.
- Breakdown happens at random and reduces your maximum capacity to 1/2 for four turns in a row
- Inventory capacity for each station is ten pieces for each type of part

The maximum allowable number of assembled parts to be transferred to the next station is determined by the dice number (as listed in Table 5-3).

TABLE 5-3 DICE TRANSLATION TABLE (ASSEMBLY STATION)

Dice number	Machine capacity
1	1
2	2
3	3
4	4
5	5
6	6

5.3.4. SALES STATION

The player at the Sales Station is responsible for delivery of the final product (Ancol, Andal, Bosscha, and Bada) to fulfill the customer demand.

Inventory capacity for this station is ten pieces of each product.

The demand amount per turn is translated using the dice translation table (as shown in Table 5-4).

TABLE 5-4 TRANSLATION TABLE (SALES STATION)

Dice Number	Demand Composition
1	0
2	1 ANCOL
3	1 ANCOL, 1 ANDAL
4	2 ANCOL, 1 ANDAL, 1 BOSSCHA, 1 BADA
5	2 ANCOL, 2 ANDAL, 1 BOSSCHA, 1 BADA
6	3 ANCOL, 2 ANDAL, 1 BOSSCHA, 1 BADA

Each fulfilled demand will be counted as revenue. On the other hand, unmet demand will be counted as backlog cost, representing reduction of customer satisfaction. Later fulfillment of unmet demand is not allowed.

5.4. Upgrade Cards

The money collected by sales can be used to improve various variables of the production system by buying Upgrade Cards. Some Upgrade Cards are valid for one station only, while others apply to the whole chain of operations.

The Upgrade Cards available for each station are:
- **Cost Saver**: This card reduces the machining cost at Production Station M1, M2, M3, and the Assembly Station for the rest of the game. Level 1 reduces the machining cost to $7, Level 2 to $4, and Level 3 to $1.
- **Capacity+**: This card increases the maximum production capacity for Production Station M1, M2, M3, and the Assembly Station for each turn. Level 1 increases the maximum production capacity for 1 part, Level 2 for 2 parts, Level 3 for 3 parts.
- **Maintenance+**: This card reduces the amount of turns that are impacted by breakdown. Level 1 reduces the breakdown to 3 turns, Level 2 to 2 turns, Level 3 to 1 turn.
- **Change Over+**: This card reduces the changeover penalty. Level 1 reduces the penalty to ¾ of the dice amount, whereas Level 2 gets rid of the changeover penalty altogether.

One Upgrade Card is available for the whole operation chain:
- **Forecast***: This card contains information corresponding to the dice numbers for one station (Purchasing, M1, M2, M3, Assembly, and Sales) for the whole game. Level 1 gives an average amount, Level 2 gives a histogram for the dice number (1-6) frequency, Level 3 gives the dice number trend for the whole game.

5.5. GAME VARIABLES

During the game, the team must monitor and consider these variables:
- **Material costs**
Material costs are calculated from the parts A, B, C, and D that are being purchased at the Purchasing Station. Part A costs $5, Part B costs $10, Part C costs $ 5, and Part D costs $ 10.

- **Machining costs**
Production Station M1, M2, M3, and the Assembly Station deliver components or products, which costs $10. Machining costs are not affected by delivered components. If a Cost Saver Card has been bought, the machining costs will be adjusted accordingly.

- **Product price**
End products are sold to customers at the following prices: Ancol = $100, Andal = $100, Bosscha = $200, Bada = $200.

- **Backlog costs**
When unmet demand exists, there will be a backlog charge, as follows: Ancol = $100, Andal = $100, Bosscha = $200, Bada = $200.

- **Upgrade cards**
Cost Saver: level 1 = $25, level 2 = $50, level 3 = $75.
Capacity+: level 1 = $1000, level 2 = $2000, level 3 = $3000
Maintenance+: level 1 = $1000, level 2 = $2000, level 3 = $3000
Changeover+: level 1 = $1000, level 2 = $2000, level 3 = $3000
Forecast: level 1 = $1000, level 2 = $2000, level 3 = $3000

5.6. RECORD FORMS

In order to help the participants monitor the previously mentioned variables, there are three types of forms, in this case based on the different stations' functions (purchasing, machining, assembling, and selling).

The Purchasing Station Form is illustrated in Figure 5-5 and focuses on monitoring costs of buying parts.

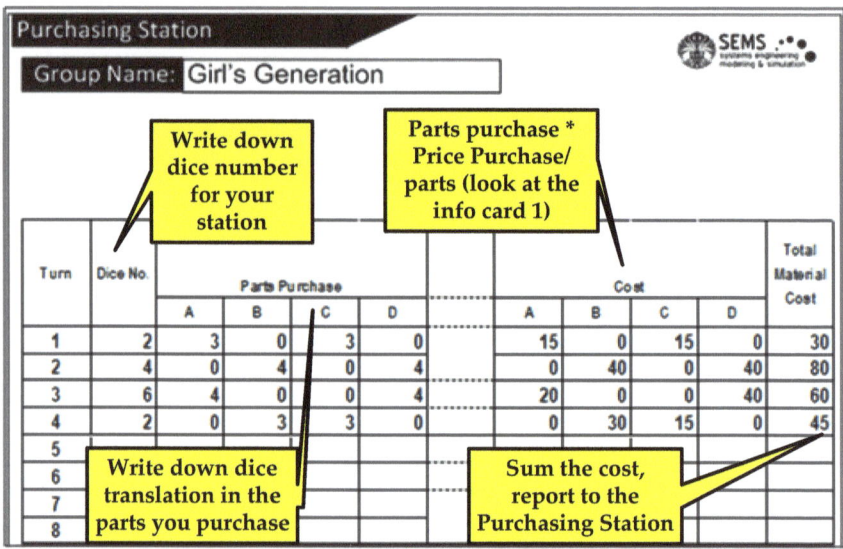

FIGURE 5-5 PURCHASING STATION RECORD FORM

At the Sales Station, where the revenue is generated, the form focuses on calculating the revenue (shown in Figure 5-6).

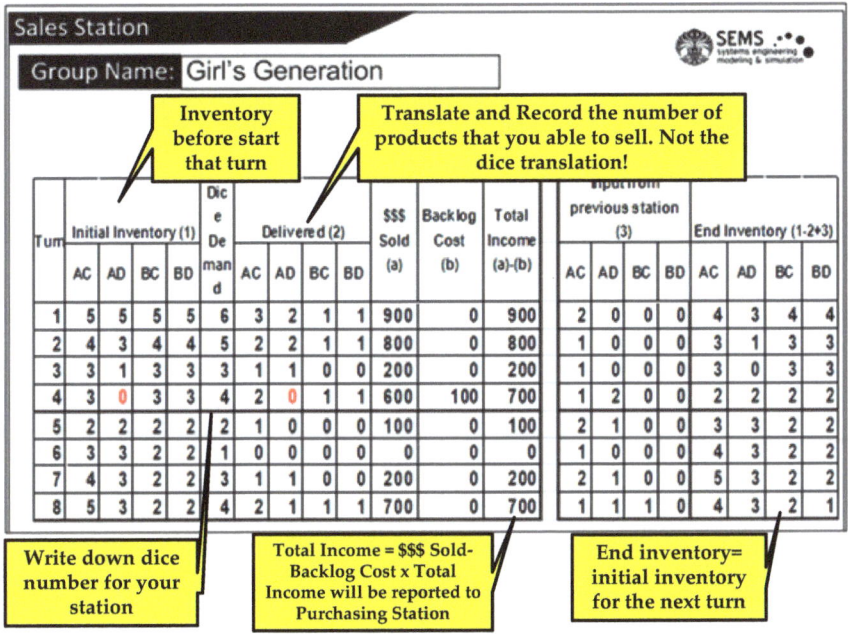

FIGURE 5-6 SALES STATION RECORD FORM

The Production Station Form is illustrated in Figure 5-7, using a step-by-step recording, moving through the columns from left to right.

The Assembly Station Form is illustrated in Figure 5-8. This is similar to the Production Station Form.

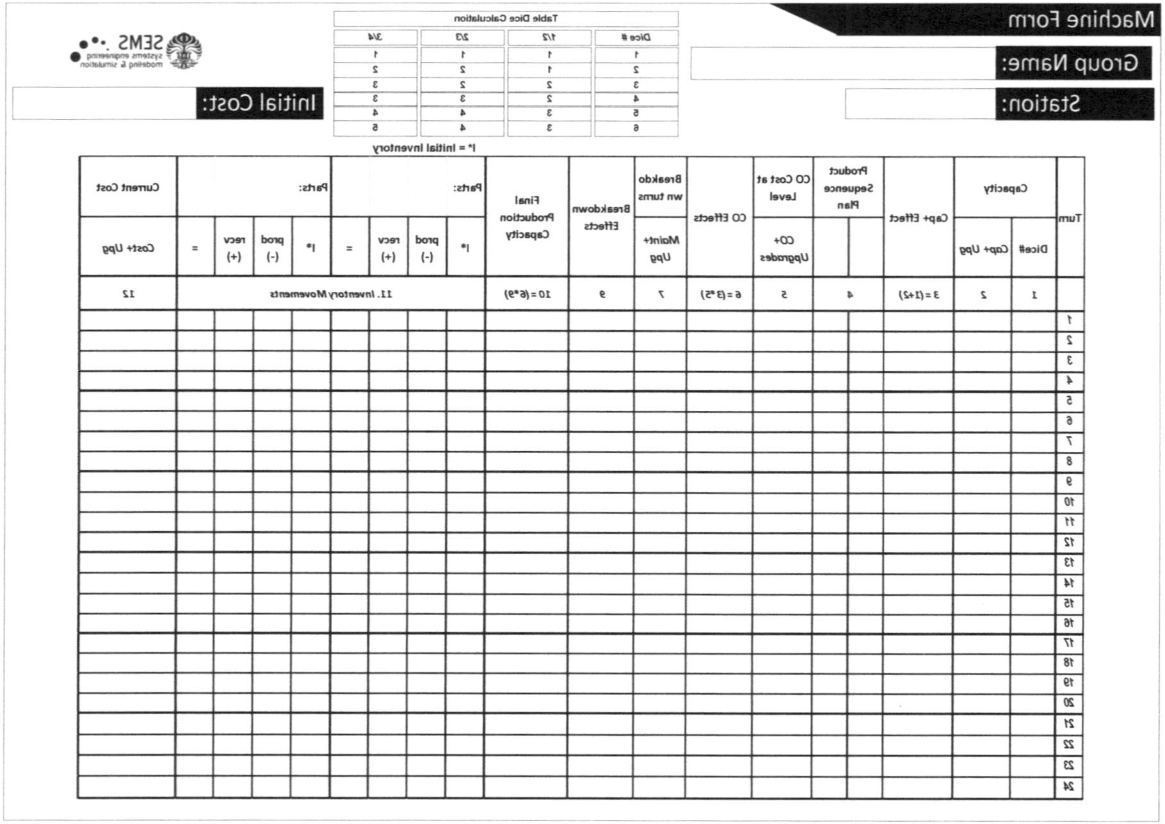

Table Dice Calculation

Dice #	1/2	2/3	3/4
1	1	1	1
2	1	2	2
3	2	2	3
4	2	3	3
5	3	4	4
6	3	4	5

I* = Initial Inventory

Turn	Capacity		Cap+ Effect	Product Sequence Plan		CO Cost at Level	CO Effects	Breakdown turns	Breakdown Effects	Final Production Capacity	Parts:				Parts:				Current Cost
	Dice#	Cap+ Upg				CO+ Upgrades		Maint+ Upg			I*	prod (-)	recv (+)	=	I*	prod (-)	recv (+)	=	Cost+ Upg
	1	2	3 = (1+2)	4		5	6 = (3*5)	7	9	10 = (6*9)	11. Inventory Movements								12
1																			
2																			
3																			
4																			
5																			
6																			
7																			
8																			
9																			
10																			
11																			
12																			
13																			
14																			
15																			
16																			
17																			
18																			
19																			
20																			
21																			
22																			
23																			
24																			

FIGURE 5-7 PRODUCTION STATION RECORD FORM

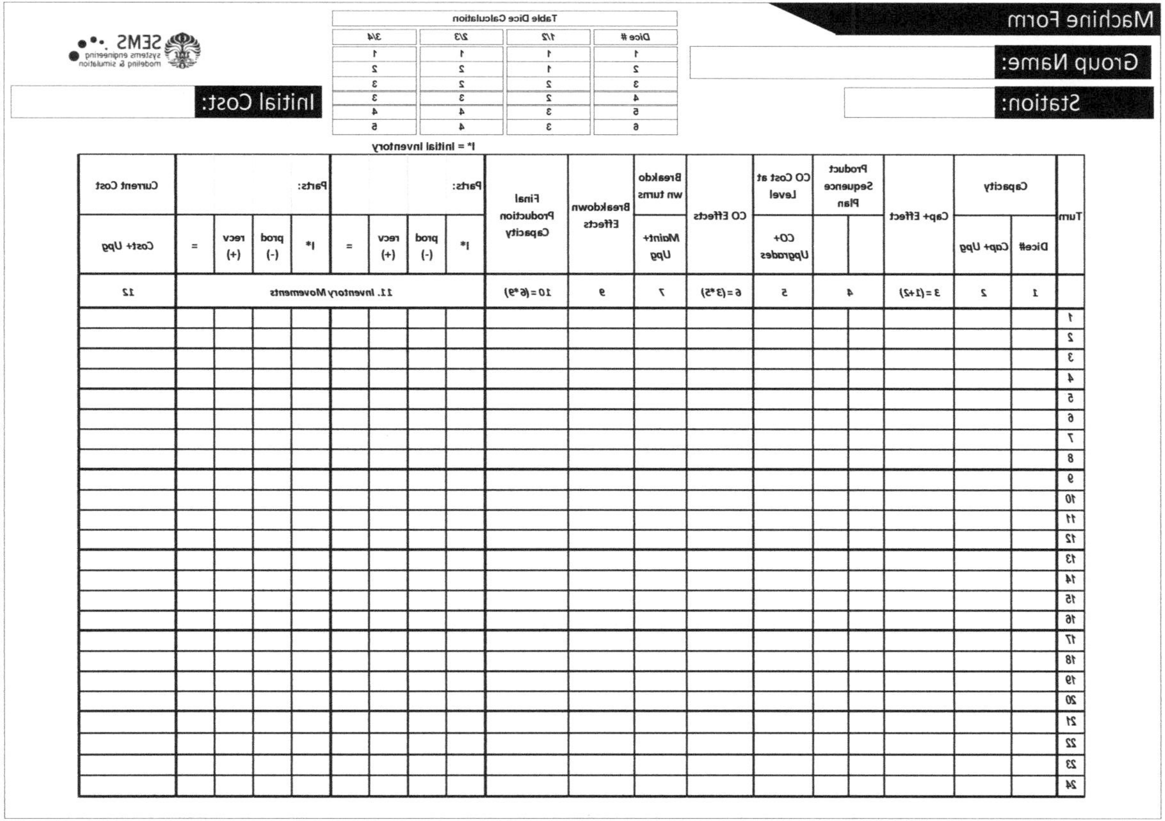

FIGURE 5-8 ASSEMBLY STATION RECORD FORM

The Company Overview Form gives an overview of the company's overall situation by monitoring the combined results of revenues, costs and upgrades. The form should be filled out by the player at the Purchasing Station, who has fewer variables to monitor than the other players.

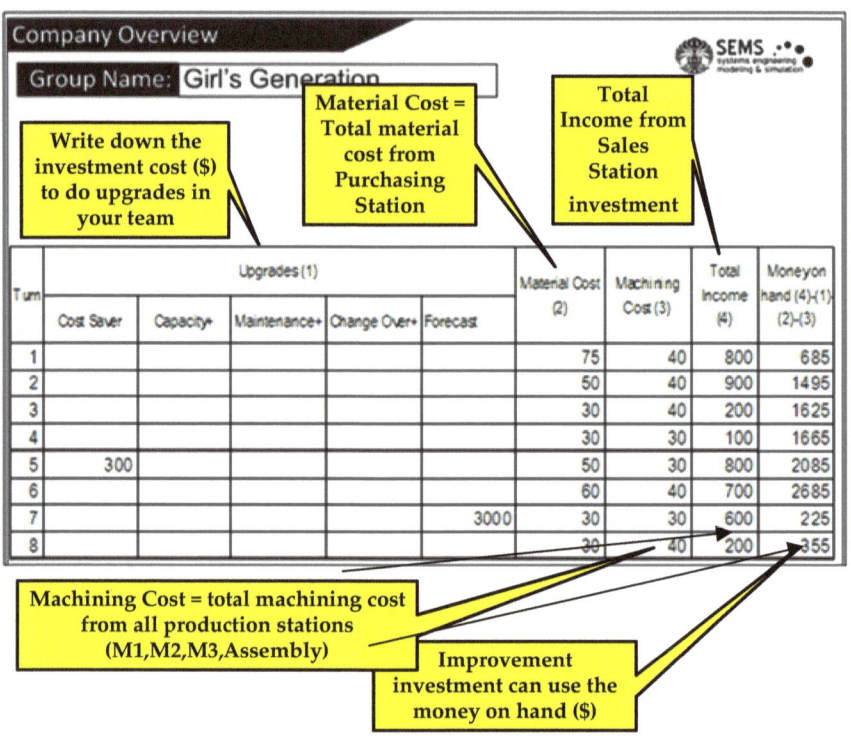

FIGURE 5-9 COMPANY OVERVIEW FORM

5.7. STAGES AND RULES FOR 6-STATIONS GAME

As mentioned in the previous chapter, OMG! involves three stages: briefing, playing and de-briefing.

5.7.1. BRIEFING

The briefing should consist of explaining the rules, the steps of the game, and a number of exercises to familiarize the players with the game, including the filling out of forms.

Before starting the game, a number of things should be arranged. If possible, each team should have at least one facilitator to monitor and explain the game rules and process. The most common mistake in turn-based games is a participant missing one or more turns. Make sure that when moving to the next cycle, all participants are at the same turn.

The rules of the game are based on the general rules for OMG! games, including:
- In this example the game is played by six people in one team. Each person represents one station.
- At the start of the game, each team is given twenty pieces of initial inventory. The initial inventory can be final products or parts, spread over all stations.
- Each team has to plan the parts and products to be produced in the next cycle. 1 cycle = 4 turns.
- All participants have to wait for the facilitator's instructions to proceed in the game.
- Investments on upgrade cards are available when the team's money on hand is enough to make the investment.

5.7.2. Playing

The basic steps of the game are as follows:
1. Decide the product's plan sequence for the next cycle. The product plan for the upcoming cycle cannot be changed until the following cycle.
2. Roll the dice or look at the number generated by the computer.
3. Translate the dice number and send parts/products to the next station's receiving buffer.
4. Record the current condition of your station on your Station Record Form.
5. Report your station's costs/income/upgrades to the player at the Purchasing Station for overall condition recording.
6. Move the parts/products in the receiving buffer to the inventory.
7. Decide on upgrades (if you have the money).
8. Report to the facilitator for upgrade investments.

If there is a facilitator, ask him to look for and note the discussion at the table. What variables do the players think are most important? Was there any significant change in strategy and when? These kinds of notes can be valuable as feedback during the debriefing.

At the end of the gameplay, the facilitator collects all forms to begin calculating the end results. The end results will be presented at the debriefing.

5.7.3. Debriefing

The debriefing should begin by asking in general about what the participants thought when making decisions during the game. Question the winning team on what they think about the priorities in decision variables, the process of selecting upgrades, and making the production plan.

The facilitator should prepare feedback questions beforehand to ensure all necessary learning points will be addressed.

At the closing of the debriefing, the facilitator can discuss the set-up of the game and check how it relates to the theoretical framework of operations management.

5.8. SPECIAL NOTES FOR PLAYING IN A MULTI-TEAM ENVIRONMENT

OMG! can be played with more than one team. Multi-team play can create a competitive environment, where the teams try to beat each other in playing. Especially if there is a 'reward' for the winning team, aside from bragging rights. The reward is not necessarily expensive. In our classes, the reward consists only of cookies and candy plus a photo-op of the winning team (which of course will be published in social media). This competitive atmosphere creates a stimulating and motivating learning environment.

In the gameplay, the multi-team environment requires the dice to be replaced with a random-number printout sheet to make sure all stations have the same variability.

READINGS

Aldrich, C. (2004). *Simulations and the future of learning : an innovative (and perhaps revolutionary) approach to e-learning.* San Francisco: Pfeiffer.

Aldrich, C. (2005). *Learning by doing : a comprehensive guide to simulations, computer games, and pedagogy in e-learning and other educational experiences.* San Francisco, CA: Pfeiffer.

Aldrich, C. (2009). *Learning online with games, simulations, and virtual worlds : strategies for online instruction* (1st ed.). San Francisco: Jossey-Bass.

Alinier, G. (2003). Nursing students' and lecturers' perspectives of objective structured clinical examination incorporating simulation. *Nurse Education Today, 23*(6), 419-426. doi: 10.1016/s0260-6917(03)00044-3

Baker, A., Navarro, E. O., & van der Hoek, A. (2003). An experimental card game for teaching software engineering. 216-223. doi: 10.1109/csee.2003.1191379

Bogost, I., & Frasca, G. (2004). *The Dean for Iowa Game and Other Rhetorical Games.* Presentation in Serious Games Summit 2004.

Cecchini, A., & Rizzi, P. (2001). *Is Urban Gaming Simulation Useful?* Simulation and Gaming, 32(4), p. 507–521.

Duke, R. D. (1974). *Gaming: the future's language.* Beverly Hills, Calif.: Sage Publications; distributed by Halsted Press.

Duke, R.D., & Geurts, J. (2004). *Policy Games for Strategic Management: Pathways into the Unknown.* Amsterdam: Dutch University Press.

Greenblat, C. S., & Duke, R. D. (1981). *Principles and practices of gaming-simulation.* Beverly Hills: Sage Publications.

Haapasalo, H., & Hyvönen, J. (2001). Simulating business and operations management–a learning environment for the electronics industry. *International Journal of Production Economics, 73*(3), 261-272. doi: 10.1016/s0925-5273(01)00088-3

Harteveld, C. (2011). *Triadic Game Design: Balancing Reality, Meaning and Play.* London: Springer-Verlag.

Houten, S.P.A. van, Scalzo, R., Bekebrede, G, Moeis, A., Bowden, N., & Mayer, I.S. (2004) *Serious Gaming for Infrastructure Design and Management: Prototypes of VENTUM on line and SIM MV2.* Proceedings of the 2004 ISAGA – SAGSAGA Conference.

Klabbers, J.H.G. (2001). *The Emerging Field of Simulation and Gaming: Meanings of a Retrospect*. Simulation and Gaming, 32(4), p. 471-480.

Klabbers, J.H.G. (2003). *Introduction to the Art and Science of Design*. Simulation and Gaming, 34(4), p. 488-494.

Leigh, E., & Kinder, J. (2001). *Fun & games for workplace learning : 40 structured learning activities to enhance workplace learning programs*. Sydney ; New York: McGraw-Hill.

Mayer, I.S. (2009). *The Gaming of Policy and the Politics of Gaming: a Review*. Simulation & Gaming, 40(6), p. 825-862.

Mayer, I.S., & Veeneman, W. (ed.) (2002). *Games in a World of Infrastructures: Simulation-Game for Research, Learning, and Intervention*. Delft: Eburon.

Mayer, I.S., Bockstael-Block, W., & Valentin, E.C. (2004). *A Building Block Approach to Simulation: An Evaluation Using Containers Adrift*. Simulation and Gaming, 35(1), p. 29-52

Neelamkavil, F. (1987). *Computer simulation and modelling*. Chichester Sussex, England ; New York: Wiley.

Proserpio, L., & Gioia, D. A. (2007). Teaching the Virtual Generation. *Academy of Management Learning & Education, 6*(1), 69-80. doi: 10.5465/amle.2007.24401703

Riis, J. O., & International Federation for Information Processing. (1995). *Simulation games and learning in production management* (1st ed.). London ; New York: Chapman & Hall on behalf of the International Federation for Information Processing.

Salen, K., & Zimmerman, E. (2003). *Rules of play : game design fundamentals*. Cambridge, Mass.: MIT Press.

Shannon, R. E. (1975). *Systems simulation : the art and science*. Englewood Cliffs, N.J.: Prentice-Hall.

Sterman, J. (2000). *Business dynamics : systems thinking and modeling for a complex world*. Boston: Irwin/McGraw-Hill.

About the Authors

Akhmad Hidayatno is the Head of Systems Engineering, Modeling and Simulation Laboratory (SEMS Lab) at Industrial Engineering Department, Universitas Indonesia. His research interest is in sustainable systems engineering, an exciting field which involves designing and managing complex systems in a sustainable way from the perspectives of economy, environment and social impacts. He hold various copyrights for various simulation games for learning, such as the kontainer game, BIEOND, franchise management simulator and other game titles. He is a system dynamics modeler, but also actively investigating how combinations of other modeling and simulation methods would be more effective in analyzing complex systems. He is a doctor in systems engineering from Universitas Indonesia.

Armand Omar Moeis holds master degree from Delft University of Technology, the Netherlands, majoring in Engineering and Policy Analysis. Prior to his graduate study, Armand gained his bachelor degree from Industrial Engineering Department, Universitas Indonesia. His current research interests are in port logistics and management using multi-method modeling approach. Beside his position as research coordinator at SEMS, Armand also holds positions in several business entities to help him to keep up his pace with real world.

Hariyanto Salim is the Chair of IPOMS, Indonesian Production and Operations Management Society. As an academic, he is actively giving lectures, trainings and workshops as well as conducting researches. As a professional, he has broad experiences in consulting services for many local and multinational companies. He is specialized in the field of Modeling, Simulation, Productivity Improvement, Operations Management, Supply Chain Management and Systems Development. He holds Master Degree in Industrial Engineering from Georgia Institute of Technology, as well as professional certificate CPIM and CSCP from APICS.

Diana Wangsa Heryanto is an Industrial Engineer from Universitas Indonesia. She was responsible to develop the detail implementation of OMG! ideas into reality.

www.ingramcontent.com/pod-product-compliance
Lightning Source LLC
Chambersburg PA
CBHW041101180526
45172CB00001B/56